The Latin American Studies Book Series

The Latin American Studies Book Series promotes quality scientific research focusing on Latin American countries. The series accepts disciplinary and interdisciplinary titles related to geographical, environmental, cultural, economic, political, urban and health-related research dedicated to Latin America. The series publishes comprehensive monographs, edited volumes and textbooks refereed by a region or country expert specialized in Latin American studies.

The series aims to raise the profile of Latin American studies, showcasing important works developed focusing on the region. It is aimed at researchers, students, and everyone interested in Latin American topics.

Submit a proposal: Proposals for the series will be considered by the Series Advisory Board. A book proposal form can be obtained from the Publisher, Andrea Brody-Barre (andrea.brody-barre@springer.com).

Verônica Garcia Donoso • Christa Reicher
Editors

Rethinking Amazonian Cities in the State of Rondônia

Springer

Editors
Verônica Garcia Donoso
Federal University of Santa Maria (UFSM)
Cachoeira do Sul, Rio Grande do Sul, Brazil

Christa Reicher
Chair of Urban Design
Institute for Urban Design and European
Urbanism RWTH Aachen University
Aachen, Germany

ISSN 2366-3421 ISSN 2366-343X (electronic)
The Latin American Studies Book Series
ISBN 978-3-032-01816-8 ISBN 978-3-032-01817-5 (eBook)
https://doi.org/10.1007/978-3-032-01817-5

The German National Library lists this publication in the German National Bibliography; detailed bibliographic data is available on the Internet at https://portal.dnb.de.

This Springer imprint is published by the registered company Springer Nature Switzerland AG
The registered company address is: Gewerbestrasse 11, 6330 Cham, Switzerland

Contents

About the Editors

Verônica Garcia Donoso is a Professor in the Architecture and Urbanism program at the Federal University of Santa Maria, Cachoeira do Sul campus, Brazil. In 2022, she was awarded the Georg Forster Research Fellowship for Experienced Researchers by the Alexander von Humboldt Foundation and conducted her research at RWTH Aachen University, Germany, from 2022 to 2024. She holds a Ph.D. (2017), a Master's degree (2012), and a Bachelor's degree (2008) in Architecture and Urbanism from the University of São Paulo, Brazil. Her research and professional experience focus on architecture and urbanism, particularly in the areas of landscape planning, ecological planning, and urban design.

Christa Reicher is the Head of the Chair of Urban Design, the Institute for Urban Design and European Urbanism, and the UNESCO Chair in Cultural Heritage and Urbanism at RWTH Aachen University, Germany. Her primary research and teaching areas include qualification strategies in urban development, urban planning, urban renewal and district development, urban and landscape design, as well as housing and residential construction.

Part I
The Challenges Currently Facing the Brazilian Amazon and the Importance of the Debate on Rondônia

Chapter 1
Introduction

Verônica Garcia Donoso and Christa Reicher

Abstract The Amazon biome is a crucial element of the world's ecosystem. Its land characteristics encompass a complex system of rivers, streams and wetlands along with a diverse array of vegetation and forest types. This complex ecosystem which is characterized by rich biodiversity, is currently facing challenges related to land use changes. The objective of this chapter is to enhance awareness of the Amazon biome by presenting information about the land characteristics of the Amazon Basin and an analysis of the most significant challenges of the region, specifically addressing the state of Rondônia, in the Brazilian Amazon. It was developed using a multi-method approach, combining bibliographic research and mapping. The results indicate that the most significant drivers of deforestation are the expansion of cattle ranching and agricultural development. Moreover, the construction of infrastructure and the arrival of settlers through colonization projects have been linked to deforestation in the state of Rondônia. The conclusions present reflections on the potential strategies for a more sustainable future.

Keywords Amazon Rainforest · Land Characteristics · Deforestation · Agribusiness

The Amazon rainforest has drawn considerable global attention and political pressure to combat deforestation in recent years, given its significant environmental impact on both local and global climates. The concern is genuine, as the Amazon

Supplementary Information The online version contains supplementary material available at https://doi.org/10.1007/978-3-032-01817-5_1.

V. G. Donoso (✉)
Federal University of Santa Maria (UFSM), Cachoeira do Sul, Rio Grande do Sul, Brazil
e-mail: veronica.donoso@ufsm.br

C. Reicher
Institute for Urban Design and European Urbanism RWTH Aachen University, Aachen, Germany

V. G. Donoso, C. Reicher (eds.), *Rethinking Amazonian Cities in the State of Rondônia*, The Latin American Studies Book Series,
https://doi.org/10.1007/978-3-032-01817-5_1

biome has lost 75 million hectares of natural vegetation from 1985 to 2021, comprising 96% of forests. This translates to an area that is a little more than twice the size of Germany.

Most deforestation took place in the Brazilian Amazon, which has lost 60.3 million hectares of forest. Urgent action is required to implement sustainable changes.

The purpose of this book is to examine the deforestation process and propose sustainable changes in specific Amazonian cities within the state of Rondônia. This state, located in the southwestern region of the Brazilian Amazon, presents a valuable case for implementing sustainable urban development due to its recent history of urbanisation and direct connection with deforestation.

The expansion of cattle ranching led to a significant decrease in natural vegetation in Rondônia. From 1985 to 2021, the area covered by natural forest vegetation decreased from 87.10% to 57.58%. In contrast, the proportion of pastureland increased from 8.21% to 36.34% during the same period.

The cities in Rondônia are relatively young, with some being less than 60 years old. Despite being situated in a biodiverse environment, their planning does not follow sustainable practises. Common issues in these cities include insufficient green areas within the urban environment, lack of sewage treatment, dense urban morphology, and inadequate infrastructure of streets and highways.

Additionally, deforestation through slash-and-burn methods presents a problem that has a negative impact on air quality. Another concern is social inequality, which is further compounded by ethnic clashes.

Rondônia has been the subject of research by Prof. Dr. Verônica Garcia Donoso as an Alexander von Humboldt Fellow, with a Georg Forster "Experienced Researcher"[1] grant from October 2022 to May 2024. The research was conducted at RWTH Aachen University in Germany under the supervision of Prof. Dr. Christa Reicher. This research project sought to examine the distinctive characteristics of urban growth and the loss of natural vegetation in the Amazon region. The project focused on the State of Rondônia and employed analytical approaches at two distinct spatial scales. The first approach employed satellite imagery to elucidate the nature and extent of changes in land use at the regional scale. In order to facilitate the contribution to the future local planning process, the second approach involved the implementation of local co-planning workshops. This book presents a comprehensive overview of the research results.

Rondônia was also topic of discussion within the international academic network "Transforming City Regions", which aims to promote sustainable design and development of regions, cities, and neighbourhoods. As part of this programme, the Chair and Institute of Urban Design at the Faculty of Architecture at RWTH Aachen University offers a Master of Science degree in Transforming City Regions (TCR).

[1] The Georg Forster Research Fellowship awards above-average qualified researchers from all disciplines from developing and emerging countries in cooperation with a German research institution. The research must address questions that are relevant to the further development of the researcher's country of origin.

This master's programme is unique in Germany, as it is taught entirely in English and attracts students from diverse international and multidisciplinary backgrounds. The curriculum emphasises project-based and design-oriented learning and incorporates numerous disciplines to address territorial challenges on a global scale.

During the summer semester of 2023, a research module was presented to students by Prof. Donoso, with Prof. Reicher supervision, which discussed various cities in the state of Rondônia. The objective of this module was to deliberate the significance of vegetation areas in the Amazon region as indispensable ecosystem services that face intense human pressures. Our research question was: if a city in the Amazon region could be redesigned according to ethno-ecological principles, what would it look like?

The course utilised GIS software for data modelling, and analysed various bibliographic sources to create maps and graphics of Rondônia state's Amazonian cities, with the aim of promoting sustainable development. The course also provided students with an overview of the Amazon region, emphasising its natural environment and ecological global significance, whilst supporting conscious urban planning of the area. Some outcomes of this semester's project were compiled into book chapters and presented in this publication.

This book is divided into three parts. The first part explores the current challenges faced by the Brazilian Amazon and highlights the significance of the debate on Rondônia. It comprises four chapters. The second chapter, titled "*Land characteristics of the Amazon region and its current challenges*", is authored by Donoso. It provides an overview of the Amazon Biome land characteristics and an analysis of the most significant challenges, particularly regarding land use Changes, with a focus on deforestation process in Rondônia. The third chapter, "*Land Use Change in the Amazon region: What is driving deforestation?*", also authored by Donoso, analyses land use and land use change in both the international and Brazilian Amazon, examining and discussing the main drivers of the deforestation. The fourth chapter, "*Layers of the Anthropocene: Occupation History of Rondônia*", authored by Donoso and Hirye, delves into the region's history, exploring various layers related to indigenous groups, quilombolas,[2] migrants, and settlers.

The second part of this book presents the findings of a research module offered in the Master Programme "Transforming City Regions" at RWTH Aachen University, Germany. This section includes four chapters, each with four case studies. Chapter 5, "*Water Sensitive Urban Design in Porto Velho*", discusses Porto Velho, the capital of the state, and seeks to develop solutions for flood adaptation through green and blue infrastructure. The authors, Singla from India and Aktürk from Turkey, focus on strategies to improve urban design and create flood-resilient drainage solutions. Chapter 6, titled "*Rethinking Amazon cities from Rondônia State: The case of Ji-Paraná*", explores sustainable development and conservation efforts in the municipality of Ji-Paraná, the second most populous city in Rondônia.

[2] Quilombolas are individuals residing in, or with a connection to, quilombos. Quilombos is the name given to the communities that were formed by black slaves who resisted the process of enslavement and escaped by forming secluded enclaves.

The authors, Saeed and Tahir Ali from Pakistan and Gül from Turkey, present practical measures to support these efforts in the urban area. Chapter 7, "*Ariquemes, the 15-Minute City: a conceptual proposal to strengthen social inclusion and sustainable greening through a 15-min urban planning system*", authored by Thakuria from India and Özdemir from Turkey, suggest implementing a planning system that promotes green living and societal inclusion. Chapter 8, "*Research on Cacoal Renewal Strategy in the context of Rondônia Rainforest*", authored by Ziaee from Iran, and Song and Bai from China, examines the development of Cacoal and proposes the sponge city concept as a solution, focusing on water absorption capacity.

This book concludes with third part, which includes two chapters. Chapter 9, "*The importance of local participatory processes in achieving sustainable changes: the experience of two ecological planning workshops in Ariquemes, Rondônia*", authored by Donoso, Covaleski and Rocha, provides insights from two participatory and multidisciplinary workshops held in Ariquemes during 2023 and 2024. These workshops were instrumental in developing local actions, projects, and strategies for sustainable development. This chapter also outlines subsequent research actions carried out following the workshops. Chapter 10, titled "*A rethink of urban planning towards and ecological planning perspective*", authored by Donoso and Reicher, offers concluding thoughts from the book's experiences.

Chapter 2
Land Characteristics of the Amazon Region and Its Current Challenges

Verônica Garcia Donoso

Abstract The Amazon biome is a crucial element of the world's ecosystem. Its land characteristics encompass a complex system of rivers, streams, and wetlands along with a diverse array of vegetation and forest types. This complex ecosystem, which is characterized by rich biodiversity, is currently facing challenges related to land use changes. The objective of this chapter is to enhance awareness of the Amazon biome by presenting information about the land characteristics of the Amazon Basin and an analysis of the most significant challenges of the region, specifically addressing the state of Rondônia, in the Brazilian Amazon. It was developed using a multi-method approach, combining bibliographic research and mapping. The results indicate that the most significant drivers of deforestation are the expansion of cattle ranching and agricultural development. Moreover, the construction of infrastructure and the arrival of settlers through colonization projects have been linked to deforestation in the state of Rondônia. The conclusions present reflections on the potential strategies for a more sustainable future.

Keywords Amazon Rainforest · Land Characteristics · Deforestation · Agribusiness

2.1 Introduction

The Amazon is the world's largest tropical forest, boasting exceptional biodiversity. It is a crucial ecosystem globally, influencing the global carbon cycle and therefore climate change as well as the global hydrological systems that support the climate and rainfall for the South American continent. The Amazon region faces the challenges of an unplanned expansion of agriculture and cattle ranching and of illegal activities related to the exploitation and extraction of natural resources, such as

V. G. Donoso (✉)
Federal University of Santa Maria (UFSM), Cachoeira do Sul, Rio Grande do Sul, Brazil
e-mail: veronica.donoso@ufsm.br

V. G. Donoso, C. Reicher (eds.), *Rethinking Amazonian Cities in the State of Rondônia*, The Latin American Studies Book Series,
https://doi.org/10.1007/978-3-032-01817-5_2

mining, timber and its products. As human activity pressure on the biome increases, a tipping point is becoming increasingly likely. This means that a situation could trigger a large-scale system change (Bottino et al., 2024; Wunderling et al., 2024; Lovejoy & Nobre, 2018).

The global magnitude of forest degradation and deforestation has a high cost for biodiversity, and its restoration is complex. Much research has been carried out on biodiversity and ecological restoration to mitigate the effects of deforestation in the Amazon region, much of it related to protected forest areas outside urban areas (Cavalcante et al., 2022; Leite-Filho et al., 2021; de Solar et al., 2016; Falcão et al., 2015; Nárgila et al., 2013). With regard to urban areas, much research was published recognizing that urban green spaces serve as habitat for wildlife and provide support for ecosystem services (Macedo et al., 2018a; LaPoint et al., 2015; Williams et al., 2015; Vergnes et al., 2013; Sushinsky et al., 2013).

Despite the acknowledged importance of forests and urban green areas, numerous questions remain regarding the optimal relationship between nature and human interference as well as the ecological planning strategies necessary to ensure high-quality living conditions for all species. This is particularly challenging in regions with complex natural ecosystems, such as the Amazon.

The objective of this chapter is to increase awareness of the Amazon biome, which has been through an extended period of time of deforestation and land degradation due to human activities such as logging, fires and agricultural practices.

It is divided into four sections. The first provides an introduction to the characteristics of the Amazon biome. The objective of this first section is to provide readers from diverse geographical and cultural backgrounds an overview of the Amazon Basin (AB) natural characteristics. Readers who are more familiar with the Amazon region may also find this overview useful. The second section will present an analysis of the most significant challenges currently facing the Amazon region. The third section will present the specific challenges facing Rondônia state, in the Brazilian Amazon. The concluding section of the chapter discusses sustainable paths for the future.

2.2 Methodology

In this chapter, an integrated multi-method approach is employed, combining particularly bibliographic research and mapping.

The Brazilian Amazon Basin maps were created using data from rivers and streams gathered by the Brazilian National Water and Basic Sanitation Agency (ANA, 2022). The topographical map analysis drew on data from the "Brazil em Relevo" initiative, which is a project of the Brazilian Agricultural Research Corporation (Embrapa). This project employs images from the Shuttle Radar Topography Mission (SRTM), collected by the National Imagery and Mapping Agency (NIMA) and the National Aeronautics and Space Administration (NASA) in 2000, to generate digital elevation models through the use of the SAR (interferometry) method. The data is accessible in raster format on the Brazilian project website (de Miranda, 2005).

To map the land use and land cover (LULC) from Rondônia state, database from the MapBiomas (2021), Collection 6.0, were used, combined with data from the Brazilian Institute of Geography and Statistics (2019, 2022).

2.3 Land Characteristics of the Amazon Region

The Amazon Rainforest covers an area of 944 million hectares, representing 47% of South America. It is shared by nine countries, with Brazil being the largest holder, accounting for 61.9% of the total area (521.9 million hectares). The Amazon Rainforest is Brazil's largest biome, covering 49.3% of the country's territory (Fig. 2.1).

The Brazilian Amazon biome is present in nine states: Acre, Amapá, Amazonas, Maranhão, Mato Grosso, Pará, Rondônia, Roraima, and Tocantins. Amazonas has the largest share of the biome, while other states also contain other biomes in their territories, such as the Cerrado and the Pantanal in Mato Grosso. Rondônia is notable for its extensive coverage of the Amazon biome, with 99% of its territory encompassed by this biome. In comparison, Mato Grosso has 54% of its territory covered by the Amazon biome, while Maranhão has 34% and Tocantins has 9%, as can be seen at Fig. 2.2 (IBGE, 2004).

In terms of natural characteristics, the Amazon Basin (AB) encompasses a wide variety of different soil types, reflecting the rich geologic diversity of origins and geomorphological processes (Quesada et al., 2011). The basin's relief is predominantly flat, with higher altitudes observed in the northern Amazonian mountains and the eastern slope of the Andes in the international Amazon region. In the Brazilian part of the Amazon region, the highest altitudes can be observed at Roraima state (Figure 2.3). The basin is characterized by a complex system of rivers, streams, and wetlands, which are irrigated by a high annual precipitation. This precipitation is distributed unevenly throughout the year, resulting in two distinct weather seasons: a rainy season and a dry season.[1] This variation in precipitation patterns has a significant impact on the hydrology, water, and soil fertility of the basin.

The AB is home to four of the world's largest river systems in terms of water discharge and drainage. Wittmann and Junk (2016) provide a comprehensive

[1] In the Brazilian Amazon, the dry season, which is defined as the period between June and November, is characterized by a monthly rainfall range of 42–162 mm. The period of rainfall is from December to May, with a monthly rainfall of 211–300 mm. The rainy season accounts for approximately 75% of the annual rainfall. A pioneering study of the two Amazonian weather seasons was conducted in 1984 (Ribeiro & Adis, 1984). In 2009, a study on the spatiotemporal variation of rainfall in the international Amazon was conducted by researchers from various institutions and published in a climatology journal. The study gathered data on precipitation from 1964 to 2003. Despite the limitations imposed by the uneven distribution of weather stations, the study contributes to the existing knowledge on rainfall in the Amazon basin, providing additional insights. These include the observation of a decline in rainfall associated with changes in the ocean and atmosphere, as well as a potential correlation with deforestation (Espinoza Villar et al., 2009).

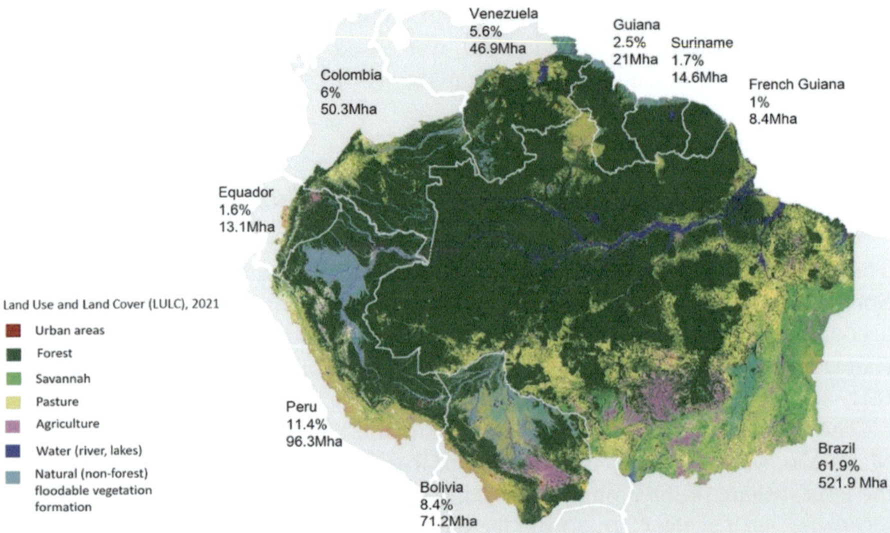

Fig. 2.1 International Amazon Biome—percentage in each country. (Source: Elaborated by Donoso, using data from Mapbiomas, 2021)

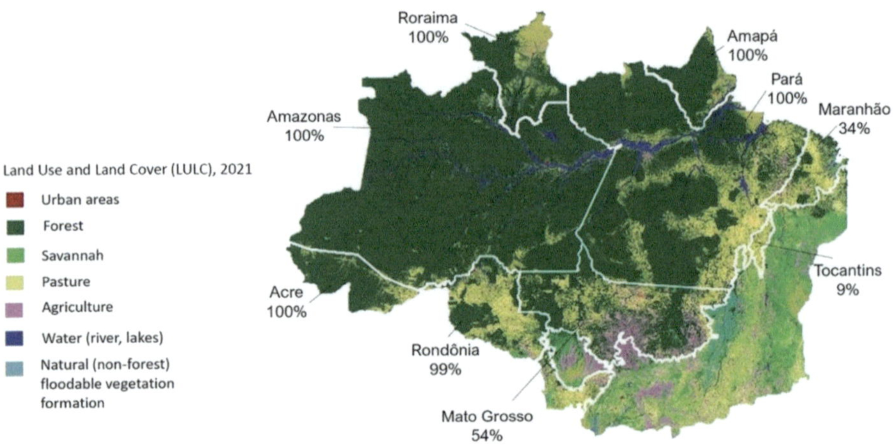

Fig. 2.2 Brazilian Amazon Biome—percentage in each state. (Source: Elaborated by Donoso, using data from Mapbiomas, 2021; IBGE, 2004)

account of the Amazon wetlands, incorporating an analysis of biodiversity. The authors describe the various types of wetlands that occur in accordance with the natural conditions, including seasonal floodplains along river systems, interfluvial wetlands of episodically flooded riparian areas along streams, permanent flooded swamps, and hydromorphic savannas. Coastal wetlands encompass mangroves and tidal floodplains. Wolfgang Junk, together with Maria Piedade, also published a book chapter about the ecology of the South American Wetland Forests, detailing the distribution of species in the Amazon (Junk & Piedade, 2010).

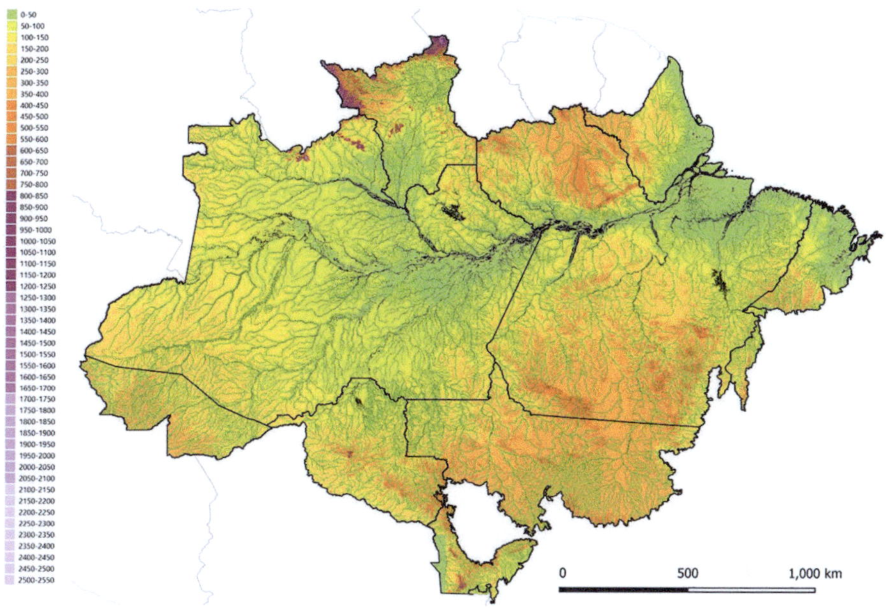

Fig. 2.3 Brazilian Amazon Basin: River system and Topography. (Source: Elaborated by Donoso, using data from ANA (2022) and de Miranda (2005))

This diversity of landscape scenarios is also present in the vegetation, and several forest types can be found in the apparently homogenous rainforest. The variations of forest types are related with the water cycle and soil, with the presence of *terra firme* forest, in non-floodable areas, and the floodable habitats or alluvial forests, known in Brazil also as *igapós, igarapés* and *várzeas* (Gonçalves et al., 2022).

Terra firme is the forest type with the highest tree alpha-diversity. In contrast, forests on *igapó*, white sand, *várzeas*, and other floodable habitats have a lower tree alpha-diversity, which is related to the poor nutrient status of the soil and the flooding periods (ter Steege et al., 2023). Additionally, other research has demonstrated the correlation between these inundation patterns and the assemblages of fauna and flora (Laranjeiras, Naka and Cohn-Haft, 2019; Barnett & Jucá, 2018; Montero et al., 2014; Parolin et al., 2013).

The biodiversity and connectivity of freshwater and terrestrial ecosystems are experiencing a significant decline due to the expansion of anthropogenic activities, including deforestation, pollution, and the construction of dams and waterways. Although the Amazon Basin is the subject of considerable scientific attention, there is still a need to extend the protection of this important ecosystem. Furthermore, it is important to recognize the importance of the freshwater ecosystem, which is a vital component of the basin (Donoso & Reicher, 2024; Castello et al., 2013).

The variation of fauna in the Amazon region is still not easy to measure by science. It is home for many endemic species, some endangered, but also many still unknown. Some identified species indicate potential to generate global human benefits, such as the case of the Fer de Lance venom, a tropical viper that help people around the world control hypertension (Lovejoy, 2019).

The Amazon Basin is a region of great cultural and social diversity, the result of a complex historical development. Upon the arrival of Europeans in the Amazon in the sixteenth century, they encountered a forest populated by numerous indigenous tribes exhibiting a rich cultural diversity. The Brazilian archaeologist Eduardo Neves (2022) asserts that the earliest known records of human habitation in the Amazon region date back approximately 14,000 years. He also estimates that approximately 5 million indigenous people inhabited the Amazon region prior to the European invasion.

The social diversity of the AB can be better understood when divided into two main categories: the forest people and traditional communities, and the urban population. This classification allows for a more nuanced understanding of the social groups that inhabit the forest. Among the first category are indigenous peoples, quilombolas, riparian communities and fisherfolk. In addition to the forest people, numerous cities and urban nuclei can be found in the Amazon Basin. This diverse social reality is the result of various historical moments, including missionary work, rubber extraction activities, mining, the promotion of settlement occupations mostly by the federal government, the creation and expansion of road infrastructures and communication infrastructure, the exploitation of water systems for energy production, the emergence of illegal activities such as logging, fires and mining, and so forth. This resulted in the social and economic development of the region, which in turn led to profound alterations in the ecosystems.

Figure 2.4 shows the topography of the Brazilian AB region, including urban areas. The map illustrates a more pronounced degree of urbanization in the coastal

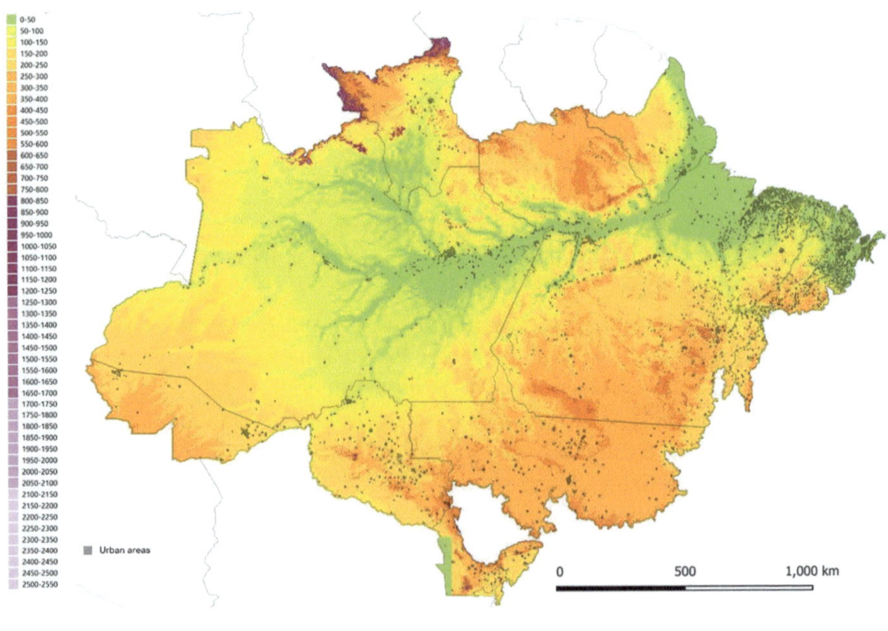

Fig. 2.4 Brazilian Amazon Basin: Topography and urban areas. (Source: Elaborated by Donoso, using data from ANA (2022), de Miranda (2005) and IBGE (2020))

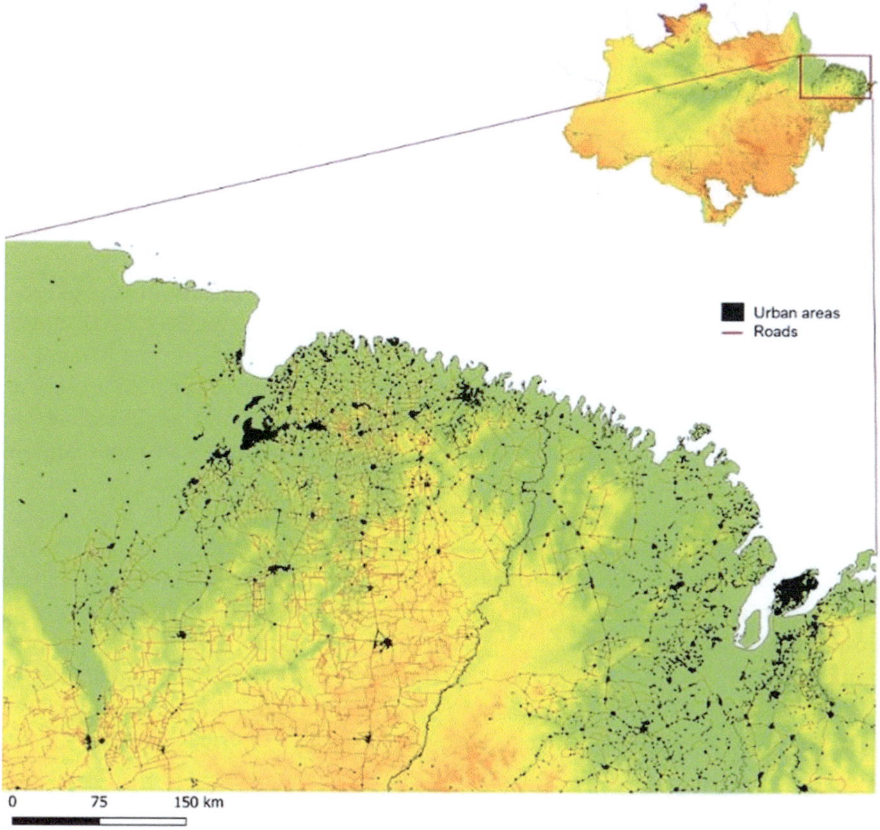

Fig. 2.5 Urban expansion of Pará and Maranhão coastal area. (Source: Elaborated by Donoso, using data from ANA (2022), de Miranda (2005) and IBGE (2020))

regions of Maranhão and Pará. This urban expansion is not a single urban agglomeration, but rather a collection of small urban nuclei dispersed across sprawled territories. This is illustrated in Figure 2.5.

2.4 The Amazon Region Current Challenges

Over the last 40 years, the Amazon has been heavily deforested and continues to deforest rapidly. From 1985 to 2020, the Amazon lost 74.6 Mha of native vegetation, with Brazil accounting for almost 60% of the deforestation (Mapbiomas, 2021).

Deforestation is one of the greatest challenges facing the Amazon rainforest, and Researchers have pointed out that deforestation is putting the Amazon in danger of reaching an irreversible tipping point. The ongoing expansion of deforestation is also affecting the local climate, intensifying drought regimes across the region (Donoso & Reicher, 2024; Anderson et al., 2018).

As a pivotal element of the Earth's climate system, the Amazon climate is intricately linked on a global scale through ocean and atmospheric circulation, as well as other interconnected systems. The interconnections between subsystems are referred to as teleconnections. For many years, researchers have been focusing on these teleconnection events, noting that the future global climate will depend on a more effective interaction between human activity and the natural environment (Wunderling et al., 2024; Kravtosov et al., 2018).

The teleconnection hypothesis offers a means of understanding how climate-related events affect other regions. For example, a significant change in sea surface elevation will result in interactions and atmospheric processes in other regions around the globe. This implies that it is not only important to protect the Amazon biome for the world's climate, but the planet must also be cared for in order to protect the Amazon. In other words, deforestation in the Amazon affects global events, and global events affect the Amazon region. This implies that the current challenges of the Amazon regions are not limited to the Amazon itself, but also related with our actions and human impacts around the globe. As an example, the potential for more frequent El Niño Southern Oscillation (ENSO) event could influence precipitation patterns in the Amazon region, leading to the emergence of extreme weather conditions such as droughts in certain regions of South America (Duque-Villegas et al., 2019).

Considering the local impacts of land use change in the Amazon region, the main drivers of deforestation are the expansion of agriculture and cattle ranching, illegal logging, fires and illegal and unregulated mining. This is a well-known problem, but it is also a political issue, with a lack of law enforcement to curb illegal activities, a lack of governance and poor decision making in the management of natural resources.

The 'Amazonia Under Pressure' atlas highlights some of the greatest challenges facing the Amazon. Published by RAISG in 2020, it points out the important role of cattle ranching in the destruction of the forest. It also focuses on the crucial role of infrastructure in driving deforestation, as the loss of vegetation is directly linked to the various access roads that cut into the forest.

Deforestation in the international Amazon is predominantly caused by anthropogenic activities, specifically pasture and agriculture, which account for 99% of the deforestation. Other anthropogenic causes, such as urbanization, only account for 1% of deforestation. Despite this, urban areas have expanded by 0.6 million hectares during the period from 1985 to 2021, representing a 152% increase. In contrast, mining activities have caused an expansion greater than 0.5%, with a 1000% increase over the same period (MapBiomas Amazônia, 2022).

About 12% of the Brazilian forest area (44 Mha) disappeared from 1985 to 2021, and most of the deforestation can be linked to cattle ranching, followed by small percentages of deforestation due to illegal mining and urban expansion. Among the Brazilian Amazon states, Pará and Rondônia stand out in the conversion of forest to agriculture or pasture (Donoso et al., 2024; Aguiar et al., 2012; Mapbiomas, 2022; Recanati et al., 2015). This book's focus on Rondônia is motivated by the necessity to discuss sustainable urban growth strategies for the region.

2.5 Rondônia Current Challenges

The state of Rondônia is positioned in the southwestern part of the Brazilian Amazon and was originally almost entirely rainforest. In 2021, compared to 1985, the natural forest vegetation area in the state reduced 66%. As shown in Figure 2.6, the natural forest vegetation was 87.10% of the state's area in 1985. In 2021, the natural forest vegetation was 57.58%. On the other hand, the pasture area, which in 1985 occupied 8.21%, increased to 36.34% in 2021 (Mapbiomas, 2021).

Rondônia state origin was the result of an international agreement between Brazil and Bolivia (Tratado de Petrópolis, Nov. 17, 1903) that allowed the construction of the Madeira Mamoré Railroad, at the beginning of the 20th century. That railroad aimed to connect Bolivia to the Atlantic Ocean to transport rubber production. This infrastructure enabled the creation of some cities, such as the capital Porto Velho in 1908 (Morelato & Cardoso, 2020).

Over the past several decades, the Brazilian State of Rondônia has been the destination of many rural migrants drawn from Brazil's middle southern regions by massive government colonization projects. Factors such as explosive population

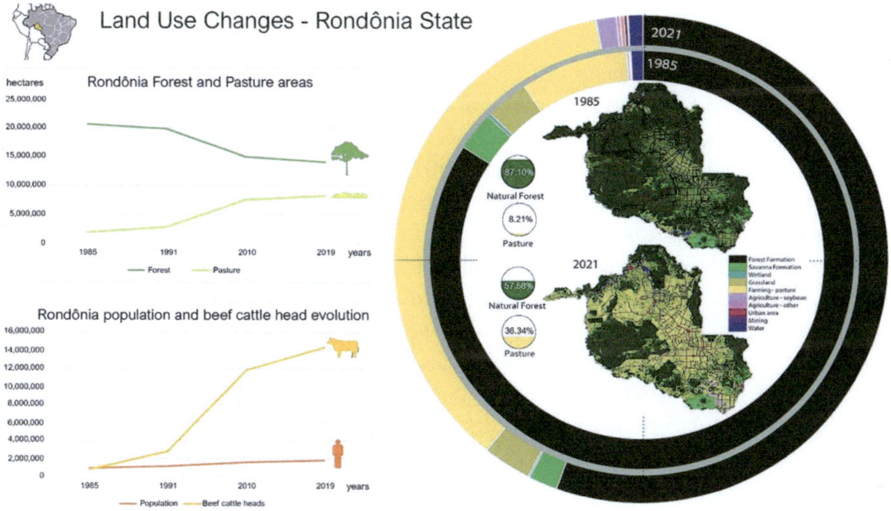

Fig. 2.6 Land use changes in Rondônia state. (Source: elaborated by Donoso. Data sources: Mapbiomas (2021), IBGE (2019, 2022))

growth, logging, mining, small scale farming, and ranching have synergistically fuelled deforestation in the state (Franco, 2019; da Cunha & Moser, 2010; de Oliveira, 2010). This state is an example of an almost irreversible tipping point of deforestation of the Amazon rainforest (Figure 2.6).

Rondônia is still undergoing a period of major agricultural expansion, currently with the growth of soy plantations.[2] Considering the urban areas, in 1985, these occupied only 0.08% of the area of Rondônia state, increasing to 0.21% in 2021.

2.6 Conclusions

It is of great importance to develop long-term, sustainable ways in which humans live and relate to nature. It thus follows that investment must be made into the development of sustainable bio-economies and the production and consumption of sustainable foods, in order to feed the increasing global population. Furthermore, it is essential to implement sustainable solutions to improve the way in which we live and plan our cities. Moreover, it is vital to enhance protection and awareness of the world's most significant biomes.

A number of environmentally friendly measures are needed. These include the establishment and management of protected areas, the creation of sustainable development plans, efficient public policies, and better planning of urban and infrastructure expansion as well as the strengthening of control and monitoring of forest areas. In fact, researchers have shown that a 1% increase in fines imposed by Brazil's Institute of the Environment and Renewable Natural Resources (IBAMA) reduced deforestation by 0.5% (Amaral, 2023b).

The importance of action at the local level cannot be underestimated. In the case of Brazil, there is a problem in controlling deforestation because permits for the suppression of native vegetation can be issued by state environmental agencies, which makes it difficult to take national and global action to control deforestation. One solution would be for states to focus on using already deforested areas to the detriment of further expansion (Gabriel, 2023).

It is important for the population to be made aware that the economic benefits of the forest outweigh those of exploiting it. In addition to agricultural productivity, regenerative agriculture and urban productivity in the Amazon can slow deforestation and stimulate regional and national development. This information is contained in the economic memorandum "A Balancing Act for Brazil's Amazonian States," published by the World Bank (Hanusch, 2023). The document also highlights that the standing forest is worth seven times more than it is exploited: the "standing" forest is estimated to be worth more than US$317 billion per year. This is the general regional and global public value for environmental protection. These values

[2] The agricultural production increased on the state with the soybean plantations from the 1990s onwards, reaching 1.34% of the state's area in 2021.

have been calculated on the basis of the ecosystem services provided by the forest, such as the generation of rainfall necessary for agriculture, soil protection against erosion and fires, carbon storage, and biodiversity conservation. On the other hand, the value of using forests for timber, mining, and agriculture generates between US$43 billion and US$98 billion annually (Amaral, 2023a).

Brazil faces the impact of colonialism and imposed policies on its development trajectories, which have resulted in marked ethno-socio-economic imbalances and marginalization (Donoso & Queiroga, 2023). In places with such mixed challenges, urban and landscape design is neither climate-resilient nor appropriate to the surrounding biodiversity (Caputo et al., 2019).

In an approach to sustainable urban development, green spaces are a key element. In combination with biodiversity, it is a step toward an ecological approach. This means recognizing that urban green spaces can contribute to the conservation of biodiversity and the principles of landscape ecology. Among other things, landscape ecology theory defines the pathway-corridor-matrix spatial pattern as a landscape component (Forman & Gordon, 1986). Together with a systemic view of the importance of open spaces (Macedo et al., 2018a, 2018b), it can support the conservation and restoration of biodiversity in urban areas.

Urban green spaces provide habitats for a wide range of wildlife. However, the size of urban areas, but also urban morphology and the availability of open spaces and green areas in cities can affect biodiversity. There are still many questions to be answered about how cities can be planned to provide good conditions for people while protecting biodiversity.

In fact, there is evidence that an increase in urban green spaces, including backyards, and functional links between them is important to avoid or mitigate the negative effects of urbanization on wildlife (Lopucki et al., 2020; Sushinsky et al., 2013; Vergnes et al., 2013; Williams et al., 2015; LaPoint et al., 2015; Matthies et al., 2017).

The importance of nature-based solutions (NbS) in contributing to sustainable development and mitigating climate change is also highlighted (Kinol et al., 2023; Sisay et al., 2023; Roggema & Tillie, 2022). Some researchers have already found that to ensure the biodiversity of native wildlife in cities, action should be taken in the early stages of urban development. They also highlight the fact that the theories of sustainable design for small cities are still inadequate (Lopucki et al., 2020).

This may not sound new or innovative to some critics, but in areas like Rondônia State, basic actions and knowledge still need to be implemented to protect the future of the Amazon region. In fact, this pathway approach could be interesting for many Global South contexts. It is important to emphasize that applications of NbS carried out in the context of the Global North provide many valuable concepts and methodologies, but they should not be readily transferred to contexts characterized by unplanned and unregulated decisions and actions, and also to contexts of extreme inequality, such as the Amazon region.Many cities in the state of Rondônia have been planned with high-density blocks and few or no urban trees. In addition, many cities have urbanized streams, known as igarapés, or have polluted them. Urban design, planning, and legal measures to increase green spaces and their connectivity in the urban fabric could support biodiversity in Amazonian cities, if combined with social and political support and environmental education.

Nevertheless, ecological planning still faces difficulties in becoming a real strategic priority in territorial planning, despite the urgency of the topic. In light of these conceptual gateways, promoting cities where there is balance between natural and human processes is every day more urgent.

Acknowledgments Research presented in this book chapter had the support of the Alexander von Humboldt Foundation, with the Georg Forster Research Fellowship for Experienced Researchers, granted for Prof. Dr. Verônica Garcia Donoso. The author gratefully acknowledges the support of the Alexander von Humboldt Foundation.

References

Aguiar, et al. (2012). Modeling the spatial and temporal heterogeneity of deforestation-driven carbon emissions: The INPE-EM framework applied to the Brazilian Amazon. *Global Change Biology, 18*, 3346. https://doi.org/10.1111/j.1365-2486.2012.02782.x

Amaral, A. (2023a). Amazônia em pé vale 7 vezes mais que exploração privada. In *Folha de São Paulo*. Caderno Ambiente. 11 de maio de 2023.

Amaral, A. (2023b). Aumento de 1% em autuações do Ibama reduz desmate em 0,5% na Amazônia, diz estudo. In *Folha de São Paulo*. Caderno Ambiente. 11 de maio de 2023.

ANA. (2022). *National water and basic sanitation agency*. Metadata Catalog da ANA—SNIRH. Retrieved May 10, 2024, from https://metadados.snirh.gov.br

Anderson, L. O., Neto, G. R., Cunha, A. P., Fonseca, M. G., et al. (2018). Vulnerability of Amazonian forests to repeated droughts. *Philosophical Transactions of the Royal Society B, 373*, 20170411. https://doi.org/10.1098/rstb.2017.0411

Barnett, A. A., & Jucá, T. (2018). Primates of igapó forests. In *Igapó (Black-water flooded forests) of the Amazon Basin* (pp. 121–133). Springer.

Bottino, M. J., Nobre, P., Giarolla, E., et al. (2024). Amazon savannization and climate change are projected to increase dry season length and temperature extremes over Brazil. *Scientific Reports, 14*, 1. https://doi.org/10.1038/s41598-024-55176-5

Caputo, S., Donoso, V. G., Izaga, F., & Britto, P. D. (2019). The democracy of green infrastructure: Some examples from Brazil and Europe. In F. Lemes de Oliveira & Mell, I. (org.) (Eds.), *Planning cities with nature. Theories, strategies and methods* (Vol. 1, 1st ed., pp. 137–152). Springer. https://doi.org/10.1007/978-3-030-01866-5_10

Castello, L., Mcgrath, D. G., Hess, L. L., et al. (2013). The vulnerability of Amazon freshwater ecosystems. *Conservation Letters, 6*(4), 217–229. https://doi.org/10.1111/conl.12008

Cavalcante, R. B. L., et al. (2022). Multicriteria approach to prioritize forest restoration areas for biodiversity conservation in the eastern Amazon. *Journal of Environmental Management, 318*, 115590. https://doi.org/10.1016/j.jenvman.2022.115590

da Cunha, E. T., & Moser, L. M. (2010). Os projetos de colonização em Rondônia. Dossiê "A construção do Sujeito Contemporâneo: Perspectiva para a Amazônia—parte II". In *Revista Labirinto—Ano X, n.14, dez.*

Miranda, EE de. (2005). *Brasil em Relevo*. Campinas: Embrapa Monitoramento por Satélite, 2005. Retrieved June, 2024, from http://www.relevobr.cnpm.embrapa.br

de Oliveira, J. L. (2010). *Rondônia: geopolítica e estrutura fundiária*. Grafiel.

de Solar, R. R. C., et al. (2016). Biodiversity consequences of land-use change and forest disturbance in the Amazon: A multi-scale assessment using ant communities. *Biological Conservation, 197*, 98–107. https://doi.org/10.1016/j.biocon.2016.03.005

Donoso, V. G., & Queiroga, E. F. (2023). Social landscape, peripheral inclusion and un-practice: Concepts for understanding social housing daily life in open spaces. *Sustainability, 15*, 12672. https://doi.org/10.3390/su151712672

Donoso, V. G., & Reicher, C. (2024). Strengthening the legal framework of protected areas in the Amazon to combat climate change. *Blue Papers, 3*(1), 9. https://doi.org/10.58981/bluepapers.2024.1.09

Donoso, V. G., Hirye, M. C. M., Gerwenat, C., & Reicher, C. (2024). Amazon deforestation and global meat consumption trends: An assessment of land use change and market data from Rondônia that shows why we should consider changing our diets. *Sustainability, 16*(11), 4526. https://doi.org/10.3390/su16114526

Duque-Villegas, M., Salazar, J., & Rendón, A. M. (2019). Tipping the ENSO into a permanent El Niño can trigger state transitions in global terrestrial ecosystems. *Earth System Dynamics, 10*, 631–650. https://doi.org/10.5194/esd-10-631-2019

Espinoza Villar, J. C., Ronchail, J., Guyot, J. L., et al. (2009). Spatio-temporal rainfall variability in the Amazon basin countries (Brazil, Peru, Bolivia, Colombia, and Ecuador). *International Journal of Climatology, 29*(11), 1574–1594. https://doi.org/10.1002/joc.1791

Falcão, J. C. F., Dáttilo, W., & Izzo, T. J. (2015). Efficiency of different planted forest in recovering biodiversity and ecological interactions in Brazili Amazon. *Forest Ecology and Management, 339*, 105–111. https://doi.org/10.1016/j.foreco.2014.12.007

Forman, R., & Gordon, M. (1986). *Landscape ecology*. John Wiley and Sons.

Franco, G. Y. (2019). A construção do "Brasil Potência": a propaganda de estímulo a migração para o Norte do Brasil—um estudo a partir do caso de Rondônia (1968-1981). *História Unisinos, 23*(1), 96–109. Retrieved March 2023, from https://www.redalyc.org/journal/5798/579865455015/html/.

Gabriel, J. (2023). Desmatamento no cerrado sobre 83% em maio e bate recorde. *Jornal Folha de São Paulo, B7*, 1. Planeta em transe, ambiente. 8 jun.

Gonçalves, A. L. S., de Oliveira, T. G., Arévalo-Sandi, A. R., et al. (2022). Composition of terrestrial mammal assemblages and their habitat use in unflooded and flooded Blackwater forests in the Central Amazon. *Peer J, 10*, e14374. https://doi.org/10.7717/peerj.14374

Hanusch, M. (2023). *A balancing act for Brazil's Amazonian states: An economic memorandum*. World Bank. http://hdl.handle.net/10986/39778

IBGE. (2004). Brazilian institute of geography and statistics. In *Mapa de Biomas do Brasil—Primeira Aproximação*. Retrieved June 10, 2024, from https://brasilemsintese.ibge.gov.br/territorio.html. IBGE.

IBGE. (2019). *Brazilian institute of geography and statistics*. Pesquisa da Pecuária Municipal. Retrieved August 23, 2023, from https://www.ibge.gov.br/estatisticas/economicas/agricultura-e-pecuaria/9107-producao-da-pecuariamunicipal.html?=&t=series-historicas

IBGE. (2022). *Brazilian institute of geography and statistics*. Population Census and Projection. Retrieved August 23, 2023, from https://cidades.ibge.gov.br/brasil/ro/pesquisa/53/49645?ano=2050

Junk, W. J., & Piedade, M. T. F. (2010). An introduction to south American wetland forests: Distribution, definitions and general characterization. In *Ecological studies* (pp. 3–25). Springer. https://doi.org/10.1007/978-90-481-8725-6_1

Kinol, A. D., Quiroga, J. A., & Kuhl, L. (2023). Opportunities for nature-based solutions to contribute to climate-resilient development pathways. *Current Opinion in Environmental Sustainability., 62*, 101297. https://doi.org/10.1016/j.cosust.2023.101297

Kravtosov, S., Grimm, C., & Gu, S. (2018). Global-scale multidecadal variability missing in state-of-the-art climate models. *NPJ Climate and Atmospheric Science, 1*, 34. https://doi.org/10.1038/s41612-018-0044-6

LaPoint, S., Balkenhol, N., Hale, J., et al. (2015). Ecological connectivity research in urban areas. *Functional Ecology, 29*, 868–878. https://doi.org/10.1111/1365-2435.12489

Laranjeiras TO, Naka, L. N., & Cohn-Haft, M. (2019). Using river color to predict Amazonian floodplain forest avifaunas in the world's largest black-water river basin. *Biotropica, 51*(3), 330–341. https://doi.org/10.1111/btp.12650

Leite-Filho, A. T., Soares-Filho, B. S., Davis, J. L., et al. (2021). Deforestation reduces rainfall and agricultural revenues in the Brazilian Amazon. *Nature Communications, 12*, 2591. https://doi.org/10.1038/s41467-021-22840-7

Lopucki, R., Klich, D., Kitowski, I., & Kersztyn, A. (2020). Urban size effect on biodiversity: The need for a conceptual framework for the implementation of urban policy for small cities. *Cities, 98*, 102590. https://doi.org/10.1016/j.cities.2019.102590

Lovejoy, T. (2019, 22 May) *Why the Amazon's biodiversity is critical for the globe: An interview with Thomas Lovejoy*. World Bank Group Feature Story. Retrieved June, 10, from https://www.worldbank.org/en/news/feature/2019/05/22/why-the-amazons-biodiversity-is-critical-for-the-globe

Lovejoy, T., & Nobre, C. (2018). Amazon tipping point. *Science Advances, 4*, eaat2340. https://doi.org/10.1126/sciadv.aat2340

Macedo, S., Queiroga, E. F., Campos, A. C. A., Galender, F., & Custodio, V. (2018a). *Os Sistemas de Espaços Livres e a Constituição da Esfera Pública Contemporânea no Brasil* (Vol. 1.ed). Editora da Universidade de São Paulo.

Macedo, S. S., CustodioV, & Donoso, V. G. (2018b). *Reflexões sobre espaços livres na forma urbana* (Vol. 1, p. 2056). FAUUSP.

Mapbiomas. (2021). *Brazil annual land use and land cover mapping project.* Collection 6.0. Retrieved Nov, 2021, from http://www.mapbiomas.org

Mapbiomas. (2022). *Em 37 anos, Amazônia perdeu 12% de florestas*. Retrieved August 23, 2023, from https://brasil.mapbiomas.org/em-37-anos-amazonia-perdeu-12-de-florestas

MapBiomas Amazônia. (2022). *Collection 4.0 of annual maps of land cover and use in the Amazon*. Retrieved November, 2023, from https://brasil.mapbiomas.org/wp-content/uploads/sites/4/2023/08/FactSheet_Panamazonia_PORTUGUES_09.08_1.pdf

Matthies, S. A., Rüter, S., Schaarschmidt, F., et al. (2017). Determinants of species richness within and across taxonomic groups in urban green spaces. *Urban Ecosystems, 20*, 897–909. https://doi.org/10.1007/s11252-017-0642-9

Montero, J. C., Piedade, M. T. F., & Wittmann, F. (2014). Floristic variation across 600 km of inundation forests (Igapó) along the Negro River, Central Amazonia. *Hydrobiologia, 729*(1), 229–246. https://doi.org/10.1007/s10750-012-1381-9

Morelato, A. A. H. N., & Cardoso, A. C. D. (2020). *Madeira Mamoré Railway Company: O legado da companytown à cidade de Porto Velho, Rondônia*. XII SIIU—Seminario Internacional de Investigação em Urbanismo.

Nárgila, G. M., et al. (2013). Avian biodiversity in multiple-use landscapes of the Brazilian Amazon. *Biological Conservation, 167*, 339–348. https://doi.org/10.1016/j.biocon.2013.08.023

Neves, E. (2022). *Sob os tempos do equinócio. Oito mil anos de história na Amazônia central*. Edusp.

Parolin, P., Wittmann, F., & Ferreira, L. V. (2013). Fruit and seed dispersal in Amazonian floodplain trees-a review. *Ecotropica, 19*(1/2), 15–32.

Quesada, C. A., Lloyd, J., Anderson, L. O., et al. (2011). Soils of Amazonia with particular reference to the RAINFOR sites. *Biogeosciences, 8*, 1415–1440. https://doi.org/10.5194/bg-8-1415-2011

RAISG. (2020). *Amazonia under pressure*. Retrieved June 10, 2024, from https://atlas2020.amazoniasocioambiental.org/en

Recanati, F., Avvievi, F., Scacabarozzi, G., et al. (2015). Global meat consumption trends and local deforestation in Madre de Dios: Assessing land use changes and other environmental impacts. *Procedia Engineering, 118*, 630–638. https://doi.org/10.1016/j.proeng.2015.08.496

Ribeiro, M. N. G., & Adis, J. (1984). Local rainfall variability-a potential bias for bioecological studies in the Central Amazon. *Acta Amazonica, 14*(1–2), 159–174. https://doi.org/10.1590/1809-43921984142174

Roggema, R., & Tillie, N. (2022). Realizing emergent ecologies: Nature-based solutions from design to implementation. *Land, 11*, 1972. https://doi.org/10.3390/land11111972

Sisay, E. D., et al. (2023). Nature-based solutions can help reduce the impact of natural hazards: A global analysis of NBS case studies. *Science of the Total Environment, 902*, 165824. https://doi.org/10.1016/j.scitotenv.2023.165824

Sushinsky, J. R., Rhodes, J. R., Possingham, H. P., Gill, T. K., & Fuller, R. A. (2013). How should we grow cities to minimize their biodiversity impacts? *Global Change Biology, 19*, 401–410. https://doi.org/10.1111/gcb.12055

ter Steege, H., Pitman, N. C. A., do Amaral, I. L., et al. (2023). Mapping density, diversity and species-richness of the Amazon tree flora. *Communications Biology, 6*, 1130. https://doi.org/10.1038/s42003-023-05514-6

Vergnes, A., Kerbiriou, C., & Clergeau, P. (2013). Ecological corridors also operate in an urban matrix: A test case with garden shrews. *Urban Ecosystems, 16*, 511–525. https://doi.org/10.1007/s11252-013-0289-0

Williams, R. L., Stafford, R., & Goodenough, A. E. (2015). Biodiversity in urban gardens: Assessing the accuracy of citizen science data on garden hedgehogs. *Urban Ecosystems, 18*, 819–833. https://doi.org/10.1007/s11252-014-0431-7

Wittmann, F., & Junk, W. J. (2016). Amazon River Basin. In C. Finlayson, G. Milton, R. Prentice, & N. Davidson (Eds.), *The Wetland book II: Distribution, description and conservation* (pp. 1–16). Springer. https://doi.org/10.1007/978-94-007-6173-5_83-2

Wunderling, N., von der Heydt, A. S., Aksenov, Y., et al. (2024). Climate tipping point interactions and cascades: A review. *Earth System Dynamics (Pembroke, Ont.), 15*(1), 41–74. https://doi.org/10.5194/esd-15-41-2024

Chapter 3
Land Use Change in the Amazon Region: What Is Driving Deforestation?

Verônica Garcia Donoso

Abstract This chapter offers insight into the Amazon region. It presents an analysis of the primary drivers of deforestation, employing land use and land cover (LULC) data and map analysis. The Mapbiomas methodology and georeferential data were used to produce LULC annual maps. The principal findings of satellite imagery analysis indicate that the conversion of natural vegetation areas to pasture and agricultural land is the primary driver of deforestation. This chapter also indicates that deforestation in the Amazon is the result of a multitude of human-related actions. It concludes by emphasizing the necessity of enhancing protection for the Amazon biome and fostering collaboration at the local level to implement sustainable planning practices.

Keywords Amazon Rainforest · Land Use and Land Cover · Deforestation · Agribusiness

3.1 Introduction

The identification of the drivers of deforestation in the world's forests has been a topic of research for decades (Pfaff, 1999; IPCC, 2021; Santos et al., 2021). In the Amazon region, deforestation has been attributed to various causes, both natural and human-related. These include forest fires, logging, and the clearing of land for agriculture and cattle farming. Other factors include the presence of infrastructure, such as roads, and governmental policy involving development projects (Donoso et al., 2024; da Silva et al., 2023; Probst et al., 2020; de Oliveira et al., 2020).

V. G. Donoso (✉)
Federal University of Santa Maria (UFSM), Cachoeira do Sul, Rio Grande do Sul, Brazil
e-mail: veronica.donoso@ufsm.br

V. G. Donoso, C. Reicher (eds.), *Rethinking Amazonian Cities in the State of Rondônia*, The Latin American Studies Book Series,
https://doi.org/10.1007/978-3-032-01817-5_3

The combination and variation of these causes of deforestation complicate efforts to address the issue and protect the biome. The deforestation of the Amazon rainforest is a matter of global concern due to the complex and interconnected nature of this ecosystem. It is estimated that the Amazon rainforest is home to over 10% of Earth's terrestrial biodiversity. In addition, evapotranspiration in the region has a cooling effect, which contributes to stabilizing Earth's climate and influencing regional rainfall patterns (Flores et al., 2024; Armstrong McKay et al., 2022; Staal et al., 2018; Barnett & Jucá, 2018; Castello et al., 2013).

The analysis of deforestation rates in the Amazon region has changed over time due to differences in methodology, data resolution, and the availability of satellite imagery. Michael Williams (2002) conducted a study of global deforestation from prehistoric times to the mid-1990s and identified factors driving deforestation throughout world history. According to his analysis, the development of agriculture beginning in the Middle Ages and the domestication and control of fire were part of the first major assault on the world's forests. Williams' theory suggests that deforestation is not a recent phenomenon, but rather, the slash-and-burn method of forest clearance is older than previously assumed, with humans having influenced forests for thousands of years.

However, Williams' data also indicates that deforestation has accelerated in the past century. Since the 1950s, land exploitation and deforestation have taken on a different scale and character, and especially since the 1980s, deforestation has drawn international attention. Initially, concerns were primarily focused on the reduction of biodiversity and vegetation loss. The accumulation of "greenhouse gases" in the atmosphere and the potential for future global warming gradually became additional concerns. Despite the international alarm, the rate of deforestation in the Global South, including the Amazon, reached its highest point in the 1980s and early 1990s (Williams, 2002).

A study conducted by Davis et al. (2020) identified a connection between tropical forest loss in the Global South and large-scale land acquisitions. The research team conducted their analysis in 15 countries across Latin America, sub-Saharan Africa, and Southeast Asia. They found that between the years 2000 and 2018, deforestation in those regions was linked to the expansion of oil palm, wood fiber, and tree plantations into native forest areas.

While methods of studying deforestation have varied over time, it is evident that the 1980s played a significant role in global deforestation, particularly in the Amazon. Until today, the rate of deforestation remains high.

In the Amazon, satellite observations indicate a decline in forest resilience, especially since the early 2000s (IPCC, 2021). Some readers may be inclined to attribute this phenomenon to population growth and urban sprawl. However, the data indicates that the primary driver of deforestation is not where or how we live, but rather the food we consume and the methods employed in its production. This suggests that, on a global scale, most forest loss is associated with the expansion of agribusiness. In the Amazon, this is particularly true for beef and soy, which are produced/raised in the Amazon and consumed in many parts of the world (Donoso et al., 2024).

This chapter explores land use and land cover data from the Amazon. The conversion of native vegetation for other purposes will be referred to as deforestation. This chapter will examine driving forces behind deforestation, focusing on the international Amazon and the Brazilian Amazon biome. It provides an essential overview for understanding deforestation in the state of Rondônia, which is the primary focus of this book.

3.2 Methods

The deforestation analysis presented here employs a mixed-method approach, combining bibliographic research, mapping, and analysis. For the mapping, the MapBiomas land use and land cover (LULC) annual maps were used. This data is freely accessible on the project's platform (Mapbiomas, 2021), and new collections are released annually by a multidisciplinary and collaborative network of researchers. This chapter draws on various collections, including 4.0, 6.0, and 8.0, which include data from 1985 to 2023. The Amazon biome international LULC data was sourced from Mapbiomas Amazônia International, with data from 2021 and 2022 contingent upon the country database.

The classification scheme for the annual maps of LULC involves four levels of classes, each with numerous categories. As an illustration of the classification scheme, the database includes examples of forest, non-forest formation, non-vegetated areas, savanna, floodable forest, wetlands, pasture, agriculture, and more. However, not all elements can be visualized at the macro scale. In this sense, the classes presented in this chapter's figures have been simplified to highlight the most prominent visible elements at each scale.

Mapbiomas employs various techniques to ensure the accuracy of its data. The Collection 8.0, which includes data from 1985 to 2023, has a global accuracy rate for legends of 84 to 90%.[1]

In addition, other data were gathered for this chapter's maps and analysis. The data from the Indigenous Territories is accessible at the National Foundation of Indigenous People's geospatial data server (FUNAI, n.d.). Additionally, data from the conservation units is accessible at the Brazilian Ministry of the Environment (MMA, n.d.).

[1] More information about the MapBiomas method and accuracy can be found at: https://brasil.mapbiomas.org/en/estatistica-de-acuracia/colecao-8/ (accessed 1 November 2023).

3.3 Land Use Change and Exploitation of the Amazon Region

3.3.1 International Amazon

The international Amazon is the area of South America originally covered by the Amazon biome. It is shared by nine countries: Brazil, Peru, Bolivia, Colombia, Venezuela, Guyana, Suriname, Ecuador, and French Guiana.

Satellite imagery analysis (Fig. 3.1) indicates that the primary drivers of deforestation in the International Amazon are related to the expansion of cattle farming and agriculture. This phenomenon is also related to international markets and the demand for food crops, meat, and timber. Furthermore, the same drivers are associated with the use of controlled and uncontrolled fires to remove natural vegetation cover. This process typically occurs after trees of commercial value have been removed, with the objective of quicky clearing land for the establishment of crops or pastures on large properties. This practice is not new: many researchers and journalists have reported it (Mapbiomas, 2023b; Gibbens, 2019). It is not only extremely dangerous for the local fauna and flora but it also contributes to greenhouse gas emissions, climate change, and a deterioration of the air quality in local communities.

From 1985 to 2020, the Amazon lost 74.6 million hectares of native vegetation. Of this total, almost 60 per cent occurred in Brazilian territory (Mapbiomas, 2021). Most deforestation in the Amazon region is the consequence of anthropogenic activities, particularly cattle ranching and agriculture, as mentioned (Fig. 3.1). These activities represent 99% of the causes of deforestation, exerting considerable pressure on the remaining forests and their watersheds. This is in accord with findings

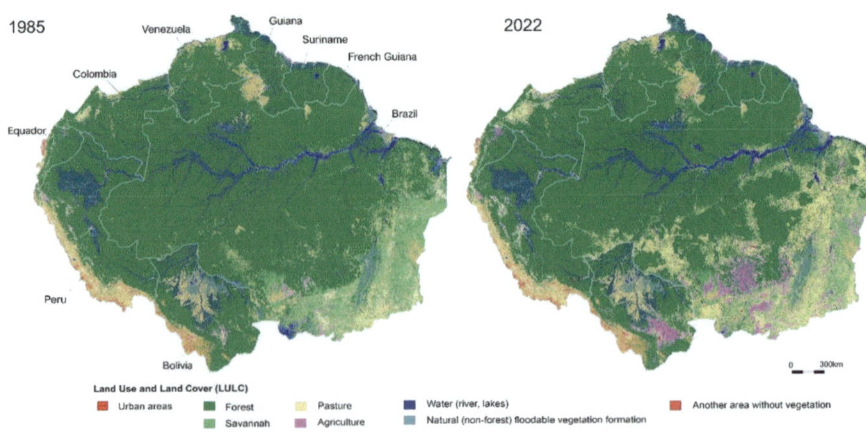

Fig. 3.1 Land Use and Land Cover in the International Amazon in 1985 and 2022. (Source: Elaborated by Donoso, using data from Mapbiomas (2022))

presented by Mapbiomas (2022), de Figueiredo et al. (2020), and Junior and Dziedzic (2021).

3.3.2 Brazilian Amazon

Researchers concur that deforestation in the Brazilian Amazon has increased since the mid-1960s, spurred by government-driven decisions made by federal government to promote occupation of the region (Hargrave & Kis-Katos, 2013; Pfaff, 1999).

To facilitate occupation, federal and state government infrastructure projects were undertaken, including the establishment of communication networks, railways, roads, and dams. The primary objective was the economic development of the region. These projects were accompanied by colonization and titling initiatives spearheaded by the military government in power from the 1960s to the 1980s. Highways and roads were constructed as part of the government's colonization projects, which were managed by the Institute of Colonization and Agrarian Reform (INCRA). The Amazon region received a steady influx of migrants, who were attracted by a government land donation program and INCRA's strategy of land distribution, while Indigenous people who had long lived in the forest were largely ignored or threatened during the process (Franco, 2019; Dall'igna, 1996; Coy, 1988).

When considering deforestation in the Brazilian Amazon biome, the states of Rondônia, Mato Grosso, and Pará stand out (Fig. 3.2). Rondônia has lost more than 34.3% of its native vegetation area, while Mato Grosso has lost 29.4%. This reflects the expansion of agribusiness, as evidenced by the case of Rondônia, which is a significant state for meat production (Wenzel et al., 2021; Wenzel & Isensee e Sá, 2018). The growth of cattle ranching operations in the Brazilian Amazon was likewise the subject of an investigation published in 2024 (Donoso et al., 2024), and the conversion of forest to pasture areas can be seen in Fig. 3.2.

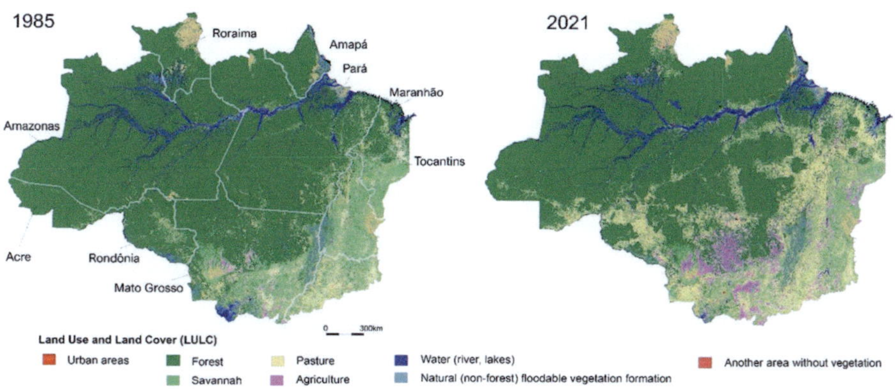

Fig. 3.2 Land Use and Land Cover in the Brazilian Amazon in 1985 and 2021. (Source: Elaborated by Donoso, using data from Mapbiomas (2022))

Mining has also been identified as a significant driver of deforestation, particularly in the Brazilian Amazon. The pathway to mining-induced deforestation is through mining companies, which may engage in leasing or off-lease activities that stimulate nearby town expansion, population growth, railway infrastructure expansion, and the conversion of permanent cropland, as evidenced by the Carajás and Trombetas mining operations in the state of Pará (Sonter et al., 2017).

Figure 3.3 shows the LULC transition diagram from 1985 to 2022 related to the Brazilian Amazon biome. In this figure, it is possible to visualize the conversion of 52 Mha of forest to agriculture during the analyzed period.

Figure 3.4 shows the total area of forest cleared prior to 2007. Additionally, it displays the Indigenous Territories ("Territórios Indígenas," TI) and the Conservation Units ("Unidades de Conservação," UC). They are territorial designations legally established by public authorities and include federal, state, district, and municipal protected areas. The Brazilian legal system of protected areas is based on a dual legal framework for the National Environmental Policy. This includes two federal laws: the Federal Forest Code (12.651/12) and the National System of Nature Conservation Units (9.985/00). UC and TI are both included in the latter (Donoso & Reicher, 2024). As illustrated in Fig. 3.4, these areas play a pivotal role in combating deforestation. It can be observed that while deforestation does occur in Conservation Units and Indigenous Territories, it is less prevalent than in unprotected areas. Figure 3.5 presents some of the key findings from the study.

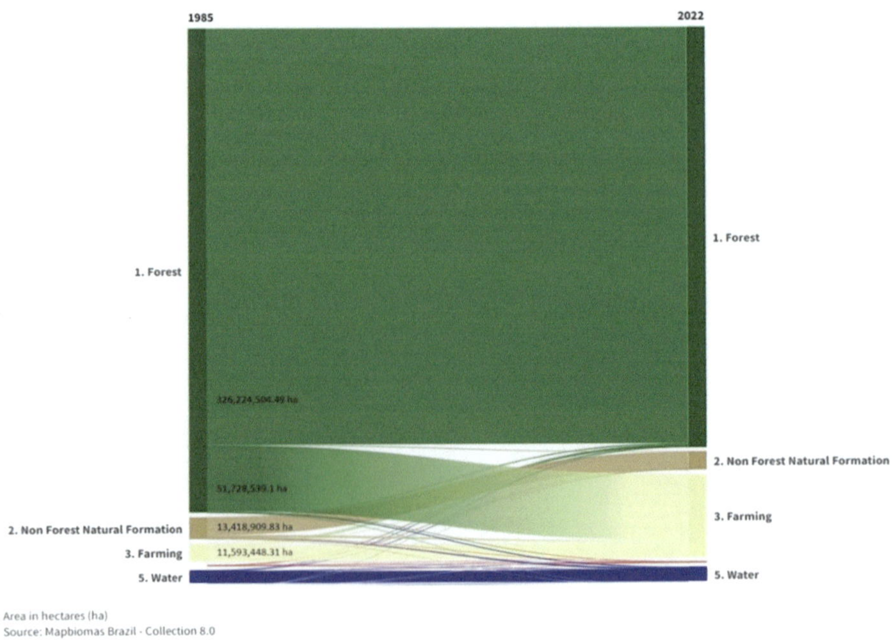

Fig. 3.3 Land Use and Land Cover transition from 1985 to 2022 at the Brazilian Amazon Biome. (Source: Elaborated by Donoso, using data from Mapbiomas, 2023a)

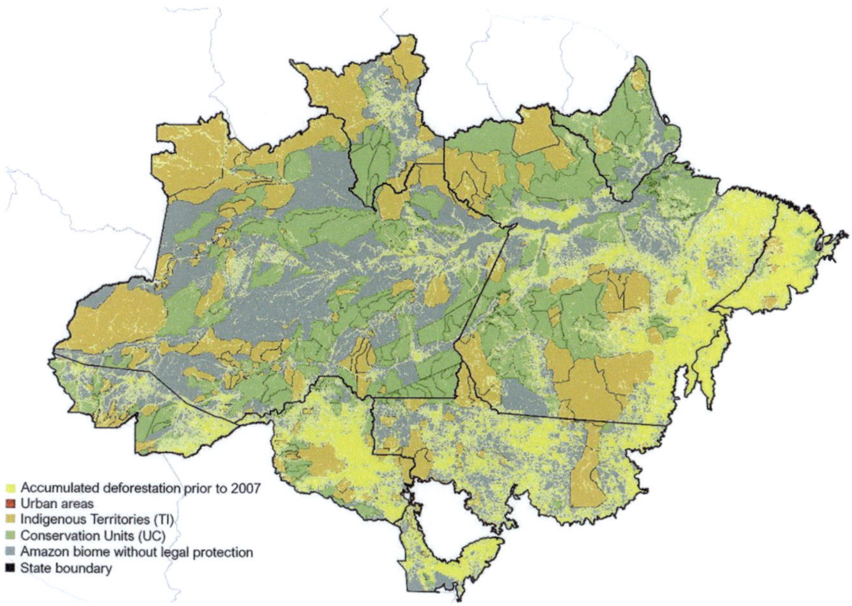

Fig. 3.4 Accumulated deforestation prior to 2007 and the important role of Indigenous Territories and Conservation Units in conserving native vegetation areas. (Source: Elaborated by Donoso, using data from Mapbiomas (2021), FUNAI, ANA (2022), de Miranda (2005), and IBGE (2020))

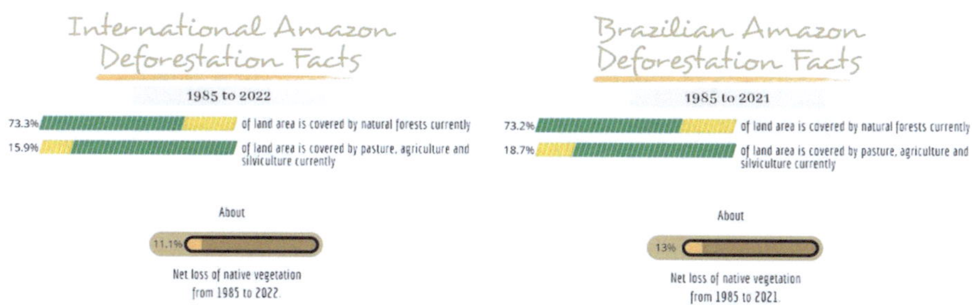

Fig. 3.5 International and Brazilian Amazon biome deforestation facts. (Source: Elaborated by Donoso, using data from Mapbiomas (2021 and 2023a))

3.4 Land Use Change and Exploitation in the States of Pará, Mato Grosso, Rondônia, and Acre

As illustrated in Figs. 3.1, 3.2, 3.3 and 3.4, the majority of land use change in the Brazilian Amazon has occurred in the states of Pará, Mato Grosso, Rondônia, and Acre. This change is primarily attributable to the expansion of agriculture and livestock production. The following Figs. 3.6, 3.7, 3.8 and 3.9, depict select examples of land use cover transition in those states.

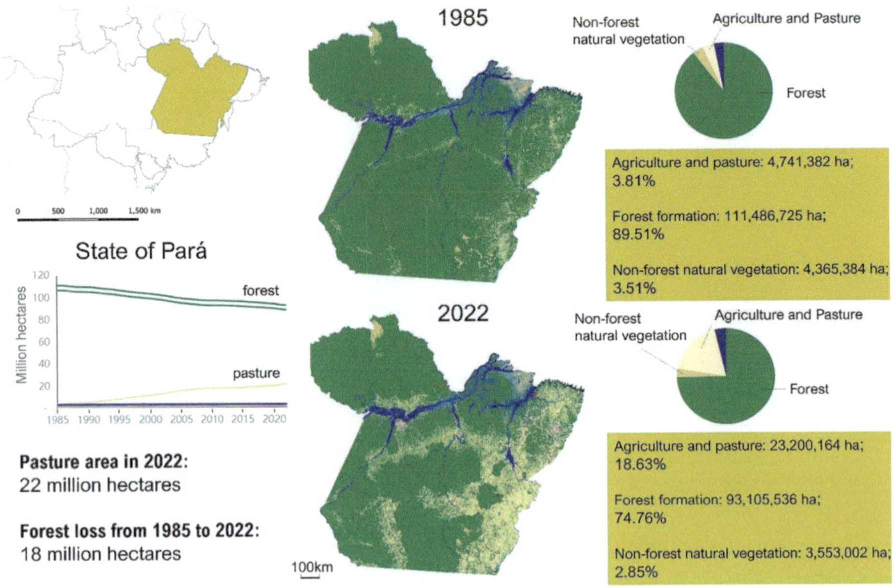

Fig. 3.6 Land Use Cover transition in the state of Pará from 1985 to 2022. (Source: Elaborated by Donoso, using data from Mapbiomas (2023a))

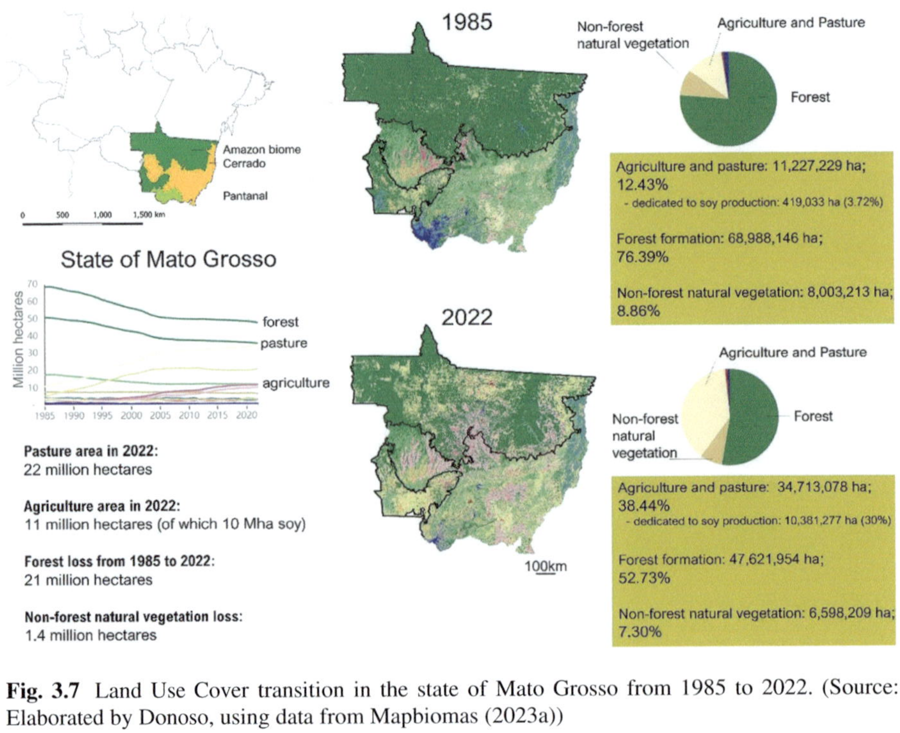

Fig. 3.7 Land Use Cover transition in the state of Mato Grosso from 1985 to 2022. (Source: Elaborated by Donoso, using data from Mapbiomas (2023a))

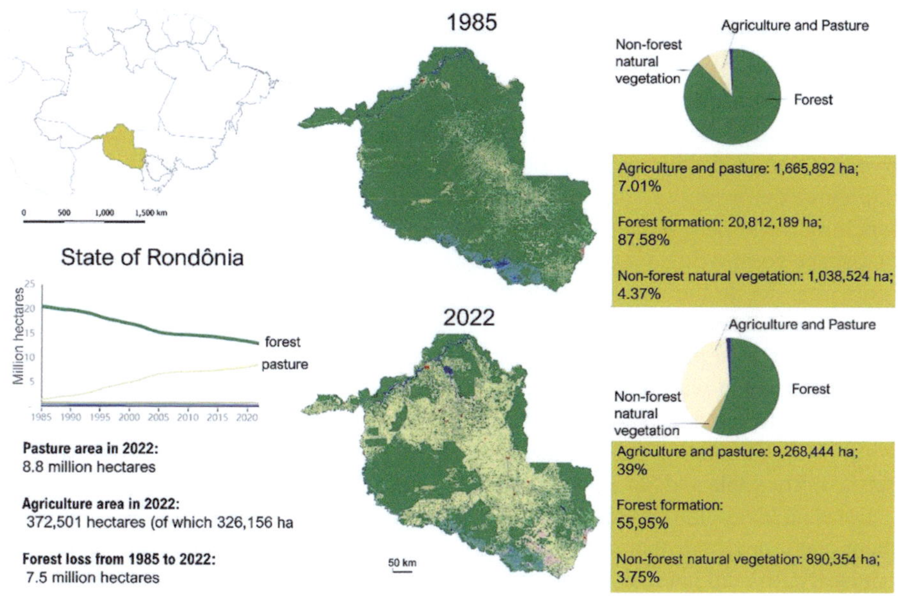

Fig. 3.8 Land Use Cover transition in the state of Rondônia from 1985 to 2022. (Source: Elaborated by Donoso, using data from Mapbiomas (2023))

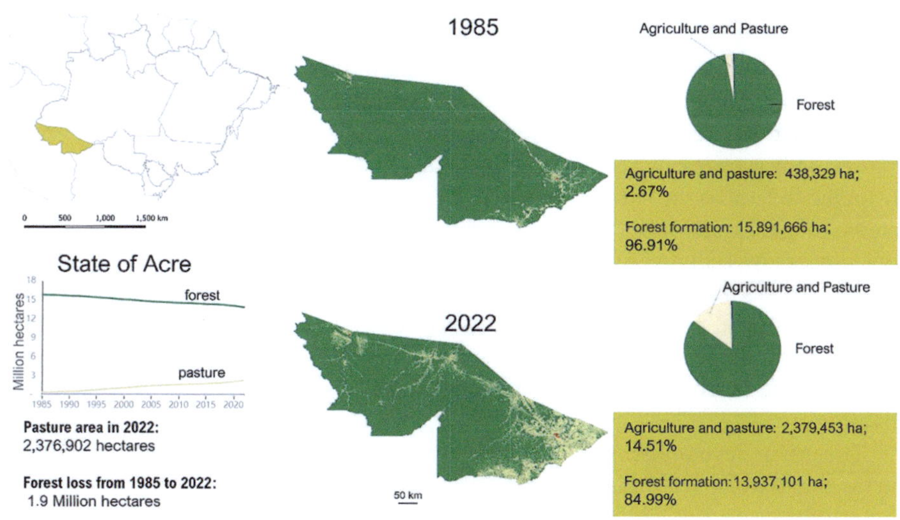

Fig. 3.9 Land Use Cover transition in the state of Acre from 1985 to 2022. (Source: Elaborated by Donoso, using data from Mapbiomas (2023))

3.5 Discussion

The satellite imagery timeframe analysis allows us to identify key periods of deforestation, and it reveals a significant degree of deforestation in the southern, southeastern, and southwestern areas of the Amazon in Brazil. This region is known as the arc of deforestation, and it is related to the expansion of agriculture and cattle ranching from the states of Pará, Mato Grosso, Rondônia, and Acre. The result is a landscape comprising a mosaic of forest fragments, with some larger areas of forest remaining due to the legal framework of protected areas, particularly Conservation Units and Indigenous Territories. This fragmentation of the forest also contributes to a less obvious process, forest degradation, which is more visible when visiting the territory than with remote sensing methods.

Many deforestation factors can be noticed in the arc of deforestation, specially related with colonization projects, the development of infrastructure and agribusiness. The states of Rondônia and Pará, for example, saw the development of infrastructure such as national highways through the forest, along with a settlement policy and federal government credit programs that subsidized agricultural loans and land redistribution carried out by INCRA, all intended to encourage greater occupation of the Amazon region.

Pará also saw a significant influx of migrants and the implementation of infrastructure projects. The establishment of a free trade zone in Manaus was also a government strategy to promote the occupation of the Amazon and the beginning of a significant deforestation trend in the rainforest. Governmental land distribution in the region was accompanied by illegal land distribution and occupation. Some migrants came not in search of land, but hoping to benefit from rubber exploitation or by finding cassiterite, gold, or trees with commercial values, such as mahogany.

Mato Grosso represents a distinct case: the colonization process commenced earlier with a series of actions, commencing in 1892, when land was transferred from the federal to the state government by the Republican Constitution. This initiated a "legal" fraud that transferred public lands to the private domain. However, this process was later regularized by colonization projects beginning in 1930 onward (1994). Mato Grosso is also unique because it includes three biomes: the Amazon, Cerrado (tropical savannah), and Pantanal (wetlands), as seen in Fig. 3.7. At present, the most deforested biome in the state is in fact the Cerrado, due to the expansion of cattle ranching and soya bean production, which is linked to the feeding of livestock.

The urban development of Rondônia began with two key infrastructure projects in the early twentieth century. The first was the construction of the Madeira-Mamoré railway between 1907 and 1912. The primary goal of this initiative was to improve the transportation network by reducing the distance between production centers and facilitating the transport of rubber. The second key development was the establishment of a telegraph network, which occurred between 1907 and 1915. Subsequent colonization projects would lead to increased migration from other parts of Brazil and changes in land use. The next chapter provides additional information about the history of Rondônia.

3.6 Conclusion

This chapter has highlighted the multiple causes of deforestation in the Amazon and shown that deforestation is primarily the result of human activities.

The deforestation of the Brazilian Amazon can be attributed to a multitude of factors. Researchers have identified the growing search for wood, agriculture, and meat products as the primary drivers of accelerated deforestation in the Amazon. Other researchers have also highlighted the infrastructure development, including road, railways, and water dams, and also land occupation programs developed by the Brazilian government dating back to the 1960s.This establishment of urban settlements by INCRA colonization projects also contributed to deforestation related to infrastructure, as the necessary connectivity between the new cities encouraged the felling of trees by illegal settlers seeking available areas and wood to exploit.

It is crucial to note that the construction of new road axes also facilitates illicit lodging and the transport of commercial wood. Satellite imagery makes it possible to identify deforestation patterns following roads, a process known as "fishbone deforestation," due to the distinctive shape of the forest paths in relation to the main road. All these drivers have encouraged the expansion of rural activities and urban settlements within the Amazon region, the exploitation of the Amazon's natural environment, and subsequent deforestation.

The satellite imagery analysis revealed the crucial role of Conservation Units and Indigenous Territories in the protection of the biome in the Brazilian territory. These areas are part of the Brazilian legal system of protected areas, specifically, the National System of Nature Conservation Units. Deforestation has primarily occurred on unprotected land, where it is illegal to conduct extractive mining and agribusiness activities. Consequently, it is imperative to expand the legal framework to encompass other territories that are currently not legally protected. Enhancing law enforcement is also crucial, along with robust collaboration at the local legal level to implement sustainable practices.

This book chapter aimed to present information about and increase awareness of deforestation in the Amazon region. It concludes by emphasizing the vital role of remote sensing technologies in producing knowledge about land use change. In conjunction with the active involvement of local communities, these technologies can support the implementation of legal sanctions and facilitate knowledge sharing and future planning.

Acknowledgments Research presented in this book chapter had the support of the Alexander von Humboldt Foundation, with the Georg Forster Research Fellowship for Experienced Researchers, granted for Prof. Dr. Verônica Garcia Donoso. The author gratefully acknowledges the support of the Alexander von Humboldt Foundation. The author also acknowledges the support of Molly Mullin in the English revision of this chapter.

References

ANA (2022) *National Water and basic sanitation agency*. Metadata Catalog da ANA—SNIRH. Retrieved May 10, 2024 from https://metadados.snirh.gov.br

Armstrong McKay, D., et al. (2022). Exceeding 1.5 °C global warming could trigger multiple climate tipping points. *Science, 377*(6611), 1–10. https://doi.org/10.1126/science.abn7950

Barnett, A. A., & Jucá, T. (2018). Primates of igapó forests. In *Igapó (Black-water flooded forests) of the Amazon Basin* (pp. 121–133). Springer.

Castello, L., Mcgrath, D. G., Hess, L. L., et al. (2013). The vulnerability of Amazon freshwater ecosystems. *Conservation Letters, 6*(4), 217–229. https://doi.org/10.1111/conl.12008

Coy, M. (1988). Desenvolvimento regional na periferia amazônica: organização do espaço, conflitos de interesses e programas de planejamento dentro de uma região de "fronteira": o caso de Rondônia. In A. Catherine (Ed.), *Fronteiras* (pp. 167–194). Universidade de Brasília; ORSTOM. ISBN 85–230–0239-1.

da Silva, R. F. B., de Castro, V. D., Nossack, F. Á., et al. (2023). Slow-down of deforestation following a Brazilian forest policy was less effective on private lands than in all conservation areas. *Communications Earth & Environment, 4*, 111. https://doi.org/10.1038/s43247-023-00783-9

Dall'igna, L. F. (1996). E o garimpo de cassiterita em Rondônia. *Revista da CPRM., 1*, 56–61.

Davis, K. F., Koo, H. I., Dell'Angelo, J., et al. (2020). Tropical Forest loss enhanced by large-scale land acquisitions. *Nature Geoscience, 13*, 482–488. https://doi.org/10.1038/s41561-020-0592-3

de Figueiredo, R. O., Cak, A., & Markewitz, D. (2020). Agricultural impacts on Hydrobiogeochemical cycling in the Amazon: Is there any solution? *Water, 12*(3), 763. https://doi.org/10.3390/w12030763

Miranda, EE de. (2005). *Brasil em Relevo.* Campinas: Embrapa Monitoramento por Satélite, 2005. Retrieved June 2024, from http://www.relevobr.cnpm.embrapa.br

de Oliveira, G., Chen, J. M., Mataveli, G. A. V., et al. (2020). Rapid recent deforestation incursion in a vulnerable indigenous land in the Brazilian Amazon and fire-driven emissions of fine particulate aerosol pollutants. *Forests, 11*(8), 829. https://doi.org/10.3390/f11080829

Donoso, V. G., & Reicher, C. (2024). Strengthening the legal framework of protected areas in the Amazon to combat climate change. *Blue Papers, 3*(1), 112. https://doi.org/10.58981/bluepapers.2024.1.09

Donoso, V. G., Hirye, M. C. M., Gerwenat, C., & Reicher, C. (2024). Amazon deforestation and global meat consumption trends: An assessment of land use change and market data from Rondônia that shows why we should consider changing our diets. *Sustainability, 16*(11), 4526. https://doi.org/10.3390/su16114526

Flores, B. M., Montoya, E., Sakschewski, B., et al. (2024). Critical transitions in the Amazon forest system. *Nature, 626*, 555–564. https://doi.org/10.1038/s41586-023-06970-0

Franco, G. Y. (2019). A construção do "Brasil Potência": a propaganda de estímulo a migração para o Norte do Brasil—um estudo a partir do caso de Rondônia (1968-1981). *História Unisinos, 23*(1), 96–109.

FUNAI. (n.d.). *National Foundation of Indigenous Peoples. Geospatial Data and Maps*. Retrieved May 10, 2024, from https://www.gov.br/funai/pt-br/atuacao/terras-indigenas/geoprocessamento-e-mapas

Gibbens, S. (2019). *The Amazon is burning at record rates—And deforestation is to blame*. National Geographic, Environment, News. August 21. Retrieved Jun 10, 2024, from https://www.nationalgeographic.com/environment/article/wildfires-in-amazon-caused-by-deforestation

Hargrave, J., & Kis-Katos, K. (2013). Economic causes of deforestation in the Brazilian Amazon: A panel data analysis for the 2000s. *Environmental and Resource Economics, 54*, 471–494. https://doi.org/10.1007/s10640-012-9610-2

IBGE. (2020). *Brazilian institute of geography and statistics*. Base de Faces de Logradouros do Brasil. Retrieved May 10, 2024, https://www.ibge.gov.br/geociencias/organizacao-do-territorio/malhas-territoriais/

IPCC. (2021). In V. Masson-Delmotte et al. (Eds.), *Climate change 2021: The physical science basis*. Cambridge Univ. Press. https://www.ipcc.ch/report/ar6/wg1/#FullReport

Junior, U. J. R., & Dziedzic, M. (2021). The water footprint of beef cattle in the Amazon region, Brazil. *Ciência Rural, 51*(8), e20190294. https://doi.org/10.1590/0103-8478cr201090294

Mapbiomas. (2021). *Brazil annual land use and land cover mapping project.* Collection 6.0. Retrieved November 2021, from http://www.mapbiomas.org

Mapbiomas. (2022). *Amazônia 2022.* Collection 4.0. Retrieved November 2023, from https://brasil.mapbiomas.org/wp-content/uploads/sites/4/2023/08/FactSheet_Panamazonia_PORTUGUES_09.08_1.pdf

Mapbiomas. (2023a). *Brazil annual land use and land cover mapping project.* Collection 8.0. Retrieved June 2024, from http://www.mapbiomas.org

Mapbiomas. (2023b). *Boletim mensal, Fevereiro 2023.* Retrieved June 2024, from https://brasil.mapbiomas.org/wp-content/uploads/sites/4/2023/09/MapBiomas_Boletim_mensal_Fevereiro2023.pdf

MMA. (n.d.). *Brazilian Environment Ministry. Georeferenced data.* Retrieved May 10, 2024, from https://antigo.mma.gov.br/areas-protegidas/cadastro-nacional-de-ucs/dados-georreferenciados.html

Pfaff, A. S. P. (1999). What drives deforestation in the Brazilian Amazon? Evidence from satellite and socioeconomic data. *Journal of Environmental Economics and Management, 37*(1), 26–43. https://doi.org/10.1006/jeem.1998.1056

Probst, B., BenYishay, A., Kontoleon, A., et al. (2020). Impacts of a large-scale titling initiative on deforestation in the Brazilian Amazon. *Nature Sustainability, 3*, 1019–1026. https://doi.org/10.1038/s41893-020-0537-2

Santos, A. M. D., Silva, C. F. A. D., Almeida Junior, P. M., et al. (2021). Deforestation drivers in the Brazilian Amazon: Assessing new spatial predictors. *Journal of Environmental Management, 294*, 113020. https://doi.org/10.1016/j.jenvman.2021.113020

Sonter, L. J., Herrera, D., Barrett, D. J., et al. (2017). Mining drives extensive deforestation in the Brazilian Amazon. *Nature Communications, 8*, 1013. https://doi.org/10.1038/s41467-017-00557-w

Staal, A., Tuinengurg, O., Bosmans, J., et al. (2018). Forest-rainfall cascades buffer against drought across the Amazon. *Nature Climate Change, 8*, 539–543. https://doi.org/10.1038/s41558-018-0177-y

Wenzel, F., Isensee e Sá, M. (2018). Sai extrativista, entra boi: a lei do mais forte em uma reserva extrativista de Rondônia. *O Eco* (4 December). Retrieved January 2024, from https://oeco.org.br/reportagens/sai-extrativista-entra-boi-a-lei-do-mais-forte-em-uma-reserva-extrativista-de-rondonia/#:~:text=Sai%20extrativista%2C%20entra%20boi%3A%20a,Rond%C3%B4nia%20%2D%20((o))eco

Wenzel, F., Hofmeister, N., Papini, P. (2021). Como os estados cuidam da Amazônia: comparamos dados de proteção, orçamento e desmatamento. *O Eco* (25 October). Accessed January 2024, from https://oeco.org.br/reportagens/como-os-estados-cuidam-da-amazonia-comparamos-dados-de-protecao-orcamento-e-desmatamento/

Williams, M. (2002). *Deforesting the earth: From prehistory to global crisis.* The University of Chicago Press.

Chapter 4
Layers of the Anthropocene: Occupation History of Rondônia

Mayumi Cursino de Moura Hirye and Verônica Garcia Donoso

Abstract This chapter provides an overview of how the territory of Rondônia state in the Brazilian Amazon region was settled. The method is bibliographic research and satellite imagery analysis. This chapter is organised into sections that divide the historical period into three so-called Anthropocene layers: original layer, the first layer, and the second layer. It concludes that the history of occupation of the region has been marked by the exploitation of natural resources in different phases of the economy, the main one being the colonisation that began in the 1960s. Finally, the Rondônia region has been affected by large-scale deforestation and the exploitation of its natural resources, such as forestry, minerals, and the expansion of agriculture. The only remaining almost unaffected areas are protected forest areas and indigenous lands.

Keywords Amazon Rainforest · Rondônia · History · Colonisation · Deforestation

4.1 Introduction

Rondônia is a state in northern Brazil with a very rich history of occupation, related to economic interests, infrastructure, and political projects for the occupation of the Amazon region. The region's history is related to indigenous groups, quilombolas, migrants, and settlers, and several disputes between these groups. Surprisingly,

M. C. de M. Hirye (✉)
Institute of the Environment and Sustainability, University of California, Los Angeles, CA, USA

Quapá Lab, Faculty of Architecture and Urbanism, University of São Paulo (USP), São Paulo, Brazil
e-mail: mhirye@ucla.edu

V. G. Donoso
Federal University of Santa Maria (UFSM), Cachoeira do Sul, Rio Grande do Sul, Brazil
e-mail: veronica.donoso@ufsm.br

V. G. Donoso, C. Reicher (eds.), *Rethinking Amazonian Cities in the State of Rondônia*, The Latin American Studies Book Series,
https://doi.org/10.1007/978-3-032-01817-5_4

these disputes are mostly from recent periods, especially from the beginning of the XX century. The history of the occupation of this territory is also marked by the exploitation of nature during different economic cycles, such as rubber and minerals, but above all by the colonisation process that brought migrants to develop the region towards agriculture and farming practices.

The administrative recognition of Rondônia as a state was only in 1982, and previously the region was a National Territory, called Guaporé, named after the Guaporé River, in the border between Brazil and Bolivia. Its history is specially related with the arrival of migrants from the beginning of the XX Century onwards (Celestino & Silva, 2015). At that time, Brazil and Bolivia signed the Treaty of Petrópolis for the construction of the Madeira-Mamoré railway, a desired infrastructure to connect the region and to transport rubber production. Also at that period, an adventurous journey to connect the Amazon region by a telegraph line was conducted by a Marshal whose name inspired in the following decades the state name, the Marshal Cândido Rondon.

In this chapter, the history of the occupation of Rondônia will be treated as the different layers of the Anthropocene. This history will be explored in three sections: firstly, the history of indigenous occupation will be addressed; together with the history of the quilombolas and the African heritage in the Amazon region; secondly, the exploitation of rubber and also the mining activities, both as economic cycles that triggered waves of migrations will be analysed; and finally, in the third section, the arrival of migrants through the Brazilian colonisation projects from the 1960s onwards will be discussed.

4.2 Original Layer: The History of Indigenous Occupation and the History of the Quilombolas and the African Heritage in the Amazon—Records from Geoglyphs and Archaeological Sites

The first records of occupation of the Amazon territory show the presence of settlements of diverse indigenous tribes approximately 14,000 years ago. The Europeans arrived at the region only in the sixteenth century, following new periods of dispute and occupation of the territory by diverse social and cultural groups.

Archaeological analyses show not only when humans were present in the Amazon rainforest but also provide insight into their way of life and social structure. In this sense, archaeological studies have had different conclusions. Archaeological exploration of the Amazon dates back to the second half of the nineteenth century, and in the 1950s, scholars began to conceptualise the economic, social, and political framework of indigenous communities living in the Amazon and other tropical forests of South America under the umbrella of tropical forest culture. These studies suggest that the development of tropical forest cultures involves a combination of local adaptation and external influences. The presence or absence of metallurgy refinements, as well as traces indicative of the cultivation of tubers like cassava and the development of river navigation techniques, have been analysed. These factors led to the distinction, in Amazonian archaeology, between floodplain and dryland environments (Neves, 2000).

In contrast, archaeological research conducted in the 1960s revealed the culture of riverside communities through ceramic remains. The so-called ceramic horizons for the forest were identified based on the ornaments and types of seasonings found within the ceramic remains of the forest (Neves, 2000).

Since the 1980s, recent investigations on the culture of communities originating in the Amazon have focused on soil studies. These studies analyse the so-called black lands, which are soils with very dark sediments that indicate fertility in archaeological sites. Although the origin of these sediments remains unclear, they may suggest long-term, uninterrupted occupancy. Further studies conducted in the 1990s examined the multilingual system of the indigenous communities and linked them to the archaeological analysis of ceramics in order to identify any connections between ethnic divisions. Despite suggestive links being discovered, the analysis does not provide definitive conclusions (Neves, 2000).

In summary, the archaeological studies do not necessarily provide definitive conclusions, but suggestions about Amazonian history of its social structures. Most of the studies provided evidence that suggest that Amazonian social formations involved the movement of people and goods across areas of the Amazon basin, and with reasonable linguistic, economic, and ecological diversity (Neves, 2000).

More recent estimations state that over 80,000 indigenous people resided within the territory of Rondônia throughout the 1930s and 40 s. Anthropologist Lévi Strauss estimated that he encountered more than 50,000 during his expedition along the Pimenta and Machado rivers, as cited in his book Tristes Tropiques, originally published in 1955 (Strauss, 1986).

Marshal Rondon's expedition not only extended the telegraph line in Rondônia but also identified and located indigenous peoples with the assistance of the Indian Protection Service (SPI). Furthermore, the SPI offered treatments for illnesses and skills training to indigenous people during the expedition and the establishment of urban centres. However, this altruistic act led to the spread of diseases amongst indigenous communities. This initial impact was followed by further conflicts due to the arrival of explorers such as rubber tappers and miners as well as migrants from colonisation projects in the 1960s. According to data, the number of identified indigenous people decreased dramatically from 80,000 in 1922 to only around 2000 in 1980 (Ferreira, 2017).

Indigenous lands of the Amazon have historically experienced multiple waves of migration at various points in the territory's history, largely due to economic cycles such as the rubber boom, which brought rubber tappers and slave migrants into the forest area. The Amazon region was marked by two centuries of slavery. Enslaved individuals who sought freedom managed to escape and sought refuge in the forest, where they established quilombos. Some of these Amazonian quilombos were discovered in the Bolivian territory of the Guaporé River Valley in approximately

1734, resulting from the expansions of São Paulo explorers seeking gold on the Guaporé River.

The discovery of gold on the Guaporé River in 1734 led to a new front of occupation. To assure the possession of the Guaporé mines, the colonial government built the Royal Fortress of Príncipe da Beira. Both for the gold extraction and for the fortress construction, the slave labour was introduced into the Guaporé valley (Cruz, 2013). In the fight for the freedom of slaves lies the origin of *quilombos*, which are settlements established by escaped slaves that can be seen in the Rondônia state today. The 1988 Brazilian Constitution gave *quilombolas* (escaped slaves or their descendants) the right to regularise their land tenure. Today there are 8 quilombos legally recognised in Rondônia State (Fundação Palmares, 2023).

Historical records suggest that the inhabitants of quilombos and the original indigenous populations had interactions. Over time, the inhabitants of quilombola communities settled in the forest and engaged in activities such as agriculture and extraction (Cruz, 2013). And their rights are safeguarded by various treaties, agreements, and declarations, emerging from 1948 and starting with the Universal Declaration of Human Rights (UN, 1948), with a emphasis on the Brazilian Constitution of 1988.

The Amazon encompasses a community of diverse ethnicities, cultures, and histories. As a result of the intricate plural scenario of forest inhabitants, they are referred to as "traditional peoples and communities" or "Forest Peoples". The forest inhabitants are indigenous people, quilombolas, riverside dwellers, rubber tappers, chestnut growers, fishermen, and others. They sustain their livelihood by hunting, fishing, traditional farming, nut harvesting, rubber tapping, and mining. The unique circumstances of each group lead to the development of distinct cultural and territorial features, including unique identities and challenges.

4.3 First Layer: Rubber Exploitation and Mining—Economic Cycles that Brought Waves of Migrants to the Region

The economic occupation and settlement of population centres in Rondônia was motivated by the exploitation of rubber at the end of the nineteenth century. It is also linked to the Brazilian-Bolivian border. Before the nineteenth century, on the Bolivian side, the Spanish explorers had not managed to penetrate the forest areas, limiting themselves to the savannah areas, known as Llanos de Mojos. Particularly suited to the savannah environment, the raising of cattle, horses, and mules sustained the settlements and remained an important economic activity in this region. In 1842, the Bolivian government created the department of Beni, covering the entire north-western part of the country, in order to encourage the colonisation of the east.

On the basis of Jesuit and colonial occupation, the exploitation of rubber began in the following century in the extensive forest areas of the region between Bolivia and Brazil (Block, 1994; Fifer, 1970). In the Brazilian Amazon, rubber was the main economic activity in the nineteenth century. Commercial rubber production began near Belém (State of Pará) between 1825 and 1850. It expanded along the Xingu and Tapajós rivers, reaching the Madeira and Purus rivers between 1850 and 1870 (de Santos, 1980). In the 1860s, rubber tappers were already observed working along the lower Purus and Madeira Rivers (Vergara, 2013). In 1878, João Gabriel de Carvalho e Melo from the state of Ceará founded the first rubber plantation in the Brazilian territory of Acre (de Pontes, 2014), which was still Bolivian territory at the time. The expansion of this activity by Brazilians in Acre territory led to negotiations between the governments of Brazil and Bolivia, which resulted in the Treaty of Petropolis, signed in 1903. The Brazilian government recognised that Bolivia had legal rights to the disputed territory and the agreement was signed with compensation to Bolivia for the definitive transfer of Acre to Brazil. Among the commitments made by Brazil were the transfer of the Brazilian rubber tappers from the left bank of the Madeira River to its confluence with the Abunã River and the construction of the Madeira-Mamoré railway, which would allow Bolivian products to flow to the Atlantic ports.

At the end of the nineteenth century, other attempts to build this railway had already failed due to the lack of specialised labour, conflicts with the indigenous communities, lack of knowledge of the physical aspects of the region, and epidemic diseases (Ferreira, 2005). In 1907, however, construction began on the "Madeira-Mamoré Railway Co.", which opened in 1912 and offered an alternative to the Madeira River and its waterfalls for the transport of rubber. The city of Porto Velho emerged from the housing nucleus for the company's employees, which emerged around the train station (IBGE, 1957). The railway also gave rise to the town of Guajará-Mirim, at the end of the line. Other towns sprang up along the 366 km railway. It is also important to note that when the railway was inaugurated in 1912, its importance had greatly diminished due to the fall in the price of rubber and the new Chilean and Argentine railways, which also transported Bolivian products (Ferreira, 2005). Currently, the BR-425 highway overlaps a large part of the Madeira-Mamoré railway route.

In 1907, the Commission for the Strategic Construction of Telegraph Lines from Mato Grosso to Amazonas was set up to connect Cuiabá, in the state of Mato Grosso, with Santo Antônio do Madeira, in the then state of Amazonas. Santo Antônio do Madeira was very close to the site of the village that would become Porto Velho. The commission was headed by Marshal Cândido Mariano da Silva Rondon, and its purpose was not only to establish communications, but also to signal the intention to definitively occupy the territory, including the creation of settlement centres (Sá, 2009). The line was inaugurated in 1915 and, like the Madeira-Mamoré Railway, it was little used, although for a different reason: its poor functioning. The original objective of extending it to Manaus was abandoned, as telegraphic technology was already becoming obsolete. However, Marshal Rondon's expedition led to the creation of several urban settlements, some of which still exist today, such as Pimenta Bueno, Ji-Paraná and Ariquemes (Browder & Godfrey, 2006), in addition to serving as a basic route for the opening of the BR-364 highway, linking Cuiabá to Porto Velho.

The decline in demand for rubber exports between 1930 and 1940 also encouraged other economic cycles, such as the collection and processing of Brazil nuts, which finally replaced rubber exploitation in 1980. In the region between Bolivia and Brazil, many communities were established on the banks of the region's navigable rivers for extractive activities, such as the Beni and Madre de Dios rivers. On the other hand, the fluctuations and changes in extractive activities also stimulated various population movements, with periods of population decline and growth.

In Rondônia, between the 1940s and 1960s, there were short economic cycles linked to the exploitation of rubber, and later cassiterite, which led to a flow of migration to the region, with the strengthening of the urban centres of Porto Velho and Ariquemes. This took place and developed in the form of manual mining, attracting a significant flow of immigrants to Rondônia. Local production already represented about half of the national amount in the early 1960s, after having reached its maximum in the 1970s with 4.7 thousand tons of tin ore, about 70% of the national amount, remaining as the main national production area until the early 1980s. Initially, exploitation was carried out in a rudimentary and individual manner, controlled by rubber tappers, who did not have legal rights to exploit. From 1971, cassiterite mining also began to be carried out industrially, remaining parallel to the mining method, in addition to mixed forms, industrial/manual. In 1987, a large deposit of cassiterite was discovered in the municipality of Ariquemes and soon became a major attractor of miners and those displaced from other industrial cassiterite ventures across the state, giving rise to Garimpo Bom Futuro. The operational framework implemented in this mine, affording miners partial control over production in conjunction with industrial exploration companies, underscores the challenges inherent in legal and industrial mining practices in the state of Rondônia (Dall'igna, 1996).

The mineral extraction movement developed further with the discovery of alluvial gold deposits in the Madeira River and in the nearest hydrographic network. This activity, which continued until recently, mobilised a large number of ferries and dredgers in a highly predatory and impactful activity that attracted a new wave of migrants.

The increase in the population in Rondônia was no longer limited to a specific area—the population and economic activities were directly linked to the water system, which was practically the only means of transport available, with the beginning of the construction of a road system that became a competitor for the establishment of new populations and productive activities. This process began in the second half of the 1950s and was planned as part of the implementation of the Amazon Economic Development Plan.

4.4 Second Layer: The Arrival of Migrants Through the Brazilian Colonisation Projects: Peaks of Deforestation Linked to the Development of Agricultural Activities

This third stage in the history of Rondônia took place between the 1960s and 1980s and marks the real intensification of the occupation of the territory, with colonisation projects that settled migrants in the region, causing urban expansion, agricultural, and livestock expansion and, consequently, deforestation.

The infrastructure development of highways and roads in the Amazon region was a turning point for deforestation and urbanisation of regions, and occurred in many states. The BR-364 highway, implemented over the axis of the telegraph line, was inaugurated in 1960, linking Rondônia to Mato Grosso state. Public and private colonisation projects were opened for small farmers as result of the land grants made by the federal government through the Institute of Colonisation and Agrarian Reform (INCRA).

In the 1970s and along the BR-364 highway, INCRA was responsible for providing land and marking territorial units in Rondônia to settle the arriving migrants. The incentives offered by the federal government attracted a new and large flow of immigrants. Agricultural activities developed rapidly, especially in the State of Rondônia, supported by the good quality of the soil and some important infrastructure works, although logistics still represented a major bottleneck for the development of the production structure. Decree-Law No. 1164 of 1971 placed a large part of the territory of Rondônia under the exclusive control of INCRA, which overlapped with an intense autonomous flow of small and medium producers who occupied vacant land. Between 1971 and 1975, five Integrated Colonisation Projects (PICs) and two Managed Settlement Projects (PADs) were implemented, all based on plots of 100 ha, with the exception of PAD Barreiro, where the plots were 250 ha (Coy, 1988).

Spatially, the settlements were concentrated along the BR-364. The area between the BR-364 axis and the banks of the Guaporé, Mamoré and Madeira rivers was protected from the intense deforestation process that preceded the occupation of the settlement. This was due to the establishment of indigenous and forest areas. Only one settlement was established on the banks of the Madeira River: Sidney Girão, in 1971, currently municipality of Nova Mamoré (Coy, 1988).

Over the course of its history, INCRA has settled just over 40,000 families in Rondônia. These settlements seldom formed economically significant urban centres, as they generally remained focused on mainly subsistence activities, with a severe lack of logistical infrastructure and technical and credit support. In this way, they survived painfully in a slow process of disintegration, later accelerated by the expansion of cattle ranching and soya farming. Nevertheless, an important component of family farming has remained in the economic base of the area, whose resistance has been strengthened by a series of support programmes, despite the persistence or even localisation of impulses that dismantle this type of social organisation of agricultural production. Despite the widespread expansion of agribusiness, the rural segment, institutionally known as family farming, in Rondônia comprises more than 75 thousand agricultural holdings and accounts for around 81% of the total of agricultural holdings, according to the last Agricultural Census (IBGE, 2017).

INCRA also manages distinct forms of settlements aimed at the sustainable exploitation of natural resources. The ‘Extractive Reserve’ (RESEX) is a type of conservation unit, recognized by state or federal law, where the areas, publicly owned, have their use granted to traditional extractive populations. These areas are intended to be “used by traditional extractive populations, whose livelihoods are based on extractivism and, complementarily, on subsistence agriculture and small-scale animal husbandry, with the basic objectives of protecting the livelihoods and culture of these populations and ensuring the sustainable use of the natural resources” (Law No. 9.985, of July 18, 2000). The recognition of RESEXs as a settlement project by INCRA facilitates the access to finance and support from the Agrarian Reform Programme. Presently, there are 25 RESEXs in Rondônia, established since 1990, housing communities engaged in rubber extraction, chestnut harvesting, and the limited-scale production of forest commodities such as copaiba oil.

Within the 1970s decade, the population grew at an impressive rate of 16% per annum, with rural population growth (18% per annum) exceeding urban population growth (14% per annum). Throughout the following decade (1980s), the growth rate halved, still the absolute number of inhabitants augmented by 642 thousand, following the completion of the BR-364 in 1984. In this decade, Rondônia state witnessed a peak in migration, when it is estimated that around 2500 families arrived in Rondônia per month (Browder & Godfrey, 2006). In order to absorb all the demand for land, INCRA adopted different settlement patterns from 1980 onwards, with plots of up to 50 hectares. Thus, during this period, INCRA settled just over 44 thousand families (Coy, 1988), which is small compared to the population contingent of one million people who arrived in Rondônia between 1970 and 1991 (Fig. 4.1).

Of the number of migrants, more than half, around 600 thousand people, settled in cities (IBGE, 2023). In this way, the opening of the agricultural frontier also modified the incipient urban network of Rondônia, which until then had been supported by Porto Velho and Guajará-Mirim, as a result of previous economic cycles. Many places were created to support the new settlements, while at the same time the places that had been telegraph stations gained dynamism. Ji-Paraná, thanks to its location on the Machado River and the support of the State at the beginning of colonisation, took the position of regional centre, becoming the second most important city in the state, after the state capital.

The colonisation projects’ cycle was marked by widespread deforestation. The first major industrial segment in the region was timber—which came to have great local expression in terms of labour absorption and production and is currently in decline and in search of sustainable development alternatives. As part of the process of opening up the agricultural frontier, there was the creation of numerous sawmills and timber industries, which established themselves in the region’s pioneering

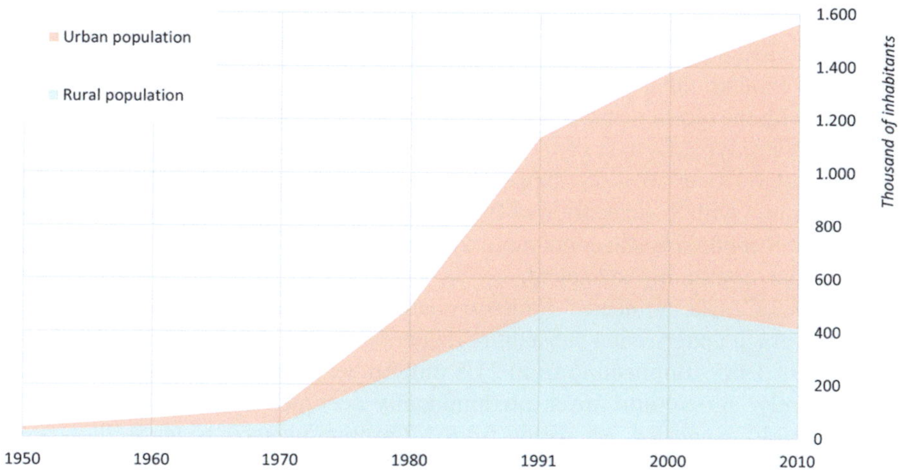

Fig. 4.1 Population in the State of Rondônia–1950 to 2010. (Source: Elaborated by Hirye, using data from IBGE, Instituto Brasileiro de Geografia e Estatística (2023, Censo Demográfico, Tabela 1288—População nos Censos Demográficos por situação do domicílio. Available at: https://sidra.ibge.gov.br/tabela/1288, accessed on 20 Nov. 2023))

frontiers. Initially, only timber of high commercial value, such as mahogany and amburana acreana, was harvested. With the opening of new roads and the advancement of the agricultural frontier, lumbering became widespread, and this sector was responsible for important exports to the Southeast, the widespread absorption of available labour, and the creation/consolidation of numerous towns.

The national and international repercussions of the strong environmental impact of urban and rural expansion in Rondônia led to the conception of the Agricultural and Forestry Plan of Rondônia (PLANAFLORO), approved in 1992.[1] This plan strengthened the approach to the management, conservation and development of natural resources, combined with socio-economic and ecological zoning, in order to reconcile development and environmental protection. In this context, a distinction was made institutionally between regions from Rondônia that could be the subject of incentives for agricultural development and those that would be the subject of incentives for conservation, with the definition of new legally protected areas (conservation units and indigenous lands). As a main result, the entire eastern part of Rondônia was defined as an area of intensification of agricultural use, while the western part was responsible for the allocation of large areas still preserved for their institutionalisation as conservation areas, with practically the only counterpart

[1] That year, the United Nations Conference on Environment and Development (UNCED), Rio-92 or Eco-92, was held in Brazil and the United Nations Framework Convention on the Environment was created. The main objective was to stabilise the concentration of greenhouse gases in the atmosphere, and the main results achieved were the establishment of Agenda 21 and an agreement called the Convention on Biological Diversity.

being the establishment of a free trade area (SUFRAMA), complemented by tax incentives (Rondônia, 1989).

In this context, the differentiation between the State's regions was accentuated. The one with the oldest occupation, corresponding to the Madeira-Guaporé mesoregion (the first dynamic axis of the economy of Rondônia) had as characteristics a situation of relative isolation, subsequently observing a broad expansion of livestock farming, with the exception of Porto Velho, which is more dynamic. The eastern part of Rondônia has become more dynamic, with agriculture, and agroindustry in particular, continuing to expand.

The LULC data delineates dramatic changes in Rondônia, wherein the initial forest coverage underwent a substantial reduction to 62% as compared to the baseline year of 1985, diminishing from 21.5 million hectares to 13.3 million hectares. Concurrently, agricultural areas, predominantly occupied by pasture, exhibited a nearly sixfold expansion, escalating from 1.7 million hectares to ten million hectares (Fig. 4.2).

In the period 2000–2010, a greater rural–urban migration was observed as a result of land concentration processes, in parallel with the increase in urban population growth rates. More recently, there has been an increase in the area devoted to crops with higher added value, in particular soya and maize, in contrast to the situation in livestock farming, as annual crops have taken over the former pastureland, which is now in decline or stagnation in several municipalities in the southeast of the state.

In the 2000s, with the installation of the Jirau and Santo Antônio hydroelectric plants, a new cycle of economic dynamism and attraction for the migrant population began: the cycle of large hydroelectric plants (total installed capacity of 6500 MW) implemented on the Madeira River. Santo Antônio, located about 10 km upstream from Porto Velho, became operational at the beginning of 2012. Jirau, 120 km upstream from Porto Velho, began operating in September 2013, reaching full

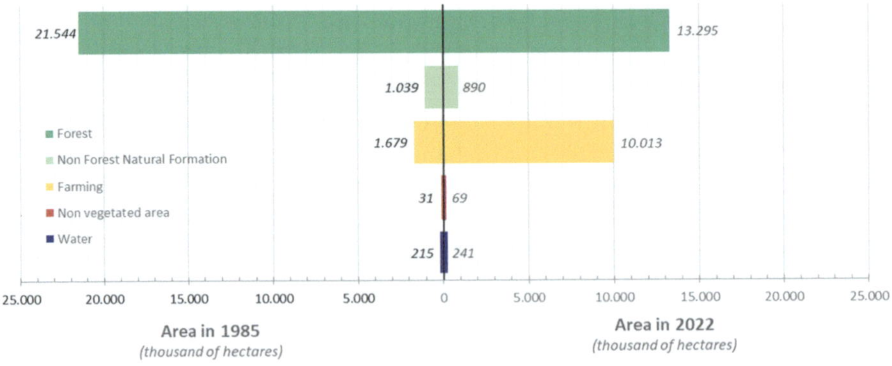

Fig. 4.2 Land use and land cover in the state of Rondônia—1985 to 2022. (Source: Elaborated by Hirye, using data from MapBiomas Project (2023) Collection 8 of the Annual Series of Land Use and Land Cover Maps of Brazil. Available at: https://brasil.mapbiomas.org/en/estatisticas/, accessed on 20 Nov. 2023)

capacity in 2016. The forced resettlement of families affected by the environmental impact of the hydroelectric plants, combined with the arrival of workers, led to an increase in urbanised areas, with the growth of the city of Porto Velho and the creation of new urban centres. The Santo Antônio Consortium built some urban nuclei: Vila Nova de Teotônio, Morrinhos, Santa Rita, Parque dos Buritis, on the right bank of the Madeira River, and Engenho Velho, São Domingos and Riacho Azul, on the left bank. In addition to these villages, several families were relocated to the INCRA Joana D'Arc I, II, and III settlements. The Jirau plant completely affected the district headquarters of Mutum Paraná. As a compensatory measure, Nova Mutum Paraná was created about 50 km north of the original site, on the banks of the BR-364 and close to the plant's construction site (ESBR, 2012).

4.5 Conclusions

Rondônia has a unique history of occupation shaped by diverse movements in the Amazon region, as highlighted by significant moments described in this chapter. These dynamics are seen in the contemporary landscape (Fig. 4.3), characterised by a juxtaposition of protected areas, wherein the presence of indigenous communities, quilombolas, and extractivists plays a pivotal role in preserving forest, and farming areas, opened by migrants and occupied predominantly by pasture. Cities are situated along the communication axis, initially following the Madeira-Mamoré railroad and subsequently gravitating towards the more dynamically evolving BR-364 axis.

Overall, the history of occupation of the region has been marked by the exploitation of natural resources during different economic periods. Additionally, the development of infrastructure projects, such as railways, telegraphs, highways, and hydroelectric plants, had various impacts on the original flora, fauna, and population throughout history.

However, the area is mostly remembered for its colonisation process, which prompted migrants to concentrate on agricultural and farming activities. This recent period of occupation had an enormous impact on the Amazon rainforest, resulting in the deforestation of forested areas, mainly due to the opening of the BR-364 highway and the arrival of settlers from colonisation projects developed by INCRA in the 1960s.

From the 2000s, in addition to the already mentioned impacts on the forest, further environmental and social issues arose due to the building of two hydroelectric power stations.

Finally, the region of Rondônia is encountering significant deforestation and the exploitation of its natural resources, including forestry, minerals, and agricultural expansion. Only the forest regions that fall under the protection of conservation units and indigenous lands remain untouched.

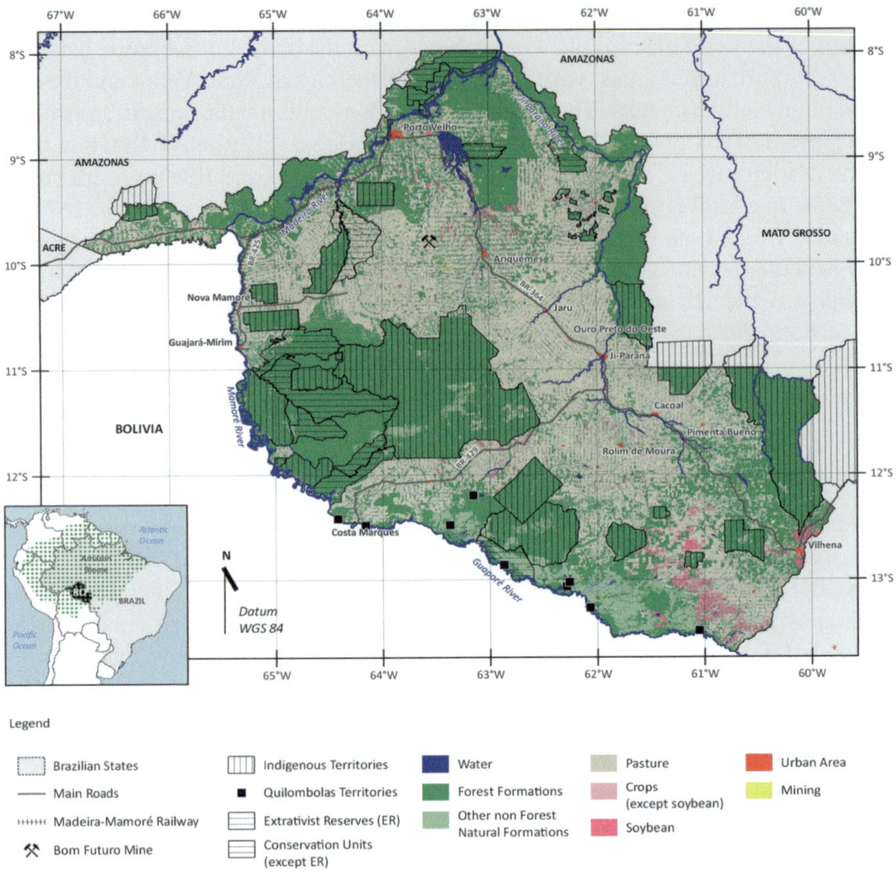

Fig. 4.3 Layers of the Anthropocene: Rondônia State in 2022. (Source: Elaborated by Hirye, using data from MapBiomas Project (2023) Collection 8 of the Annual Series of Land Use and Land Cover Maps of Brazil. Available at: https://brasil.mapbiomas.org/en/estatisticas/, accessed on 20 Nov. 2023; IBGE, Instituto Brasileiro de Geografia e Estatística (2019) Base cartográfica contínua do Brasil ao milionésimo: BCIM. Available at: https://www.ibge.gov.br/geociencias/cartas-e-mapas/bases-cartograficas-continuas/15759-brasil.html, accessed on 28 Nov. 2023 and FUNAI Fundação Nacional do Índio (2023). Terras Indígenas do Brasil: dados geoespaciais. Available at: https://www.gov.br/funai/pt-br/atuacao/terras-indigenas/geoprocessamento-e-mapas, accessed on 28 Nov. 2023)

Acknowledgement Research presented in this book chapter had the support of the Alexander von Humboldt Foundation, with the Georg Forster Research Fellowship for Experienced Researchers, granted for Prof. Dr. Verônica Garcia Donoso. The Authors would also like to extend their gratitude to the LabQUAPÁ, from the University of São Paulo (Brazil), and the MapBiomas Brazil project, for their support in facilitating the research visits of the first author to Rondônia state.

References

Block, D. (1994). *Mission culture on the upper Amazon: Native tradition, Jesuit enterprise & secular policy in Moxos, 1660–1880*. University of Nebraska Press.

Browder, J. O., & Godfrey, B. J. (2006). In Trad. Gisele Vieira Goldstein con colab & d. J. Souza (Eds.), *Rainforest cities: Urbanization, development, and globalization of the Brazilian Amazon*. Editora de la Universidad Federal de Amazonas.

Celestino, G. C. S., & Silva, J. E. (2015). Populações Guaporeanas: configuração e reconfigurações no mundo neocolonial. *Revista Veredas Amazônicas., 4*, 1.

Coy, M. (1988). Desenvolvimento regional na periferia amazônica: organização do espaço, conflitos de interesses e programas de planejamento dentro de uma região de "fronteira": o caso de Rondônia. In A. Catherine & B. Becker (Eds.), *Fronteiras* (pp. 167–194). Universidade de Brasília; ORSTOM. ISBN 85-230-0239-1.

Cruz, T. (2013). O processo de formação das comunidades quilombolas do Vale do Guaporé. In *Anais XXVII Simpósio Nacional de História*.

Dall'igna, L. F. (1996). E o garimpo de cassiterita em Rondônia. *Revista da CPRM, 1*, 56–61.

de Pontes, C. J. F. (2014). O primeiro ciclo da borracha no Acre: da formação dos seringais ao grande colapso. *South American Journal of Basic Education, Technical and Technological, 1*(1), 107–103.

de Santos, R. A. O. (1980). *História econômica da Amazônia: 1800–1920*. T. A. Queiroz.

ESBR, Energia Sustentável do Brasil. (2012). Usina Hidrelétrica Jirau—Relatório Final de Implantação. In *Programa de Remanejamento das Populações Atingidas. Anexo ao Relatório Final de Implantação* (p. 75).

Ferreira, M. R. (2005). *A ferrovia do diabo*. Melhoramentos.

Ferreira, IA. (2017). *Povos Indígenas de Rondônia: rumo aos 518 anos de resistência*. Teoria e Debate, ed. 167, 18 dez. 2017. Retrieved October 11, 2023, from https://teoriaedebate.org.br/2017/12/18/povos-indigenas-rondonia-rumo-518-anos-de-resistencia/

Fifer, J. V. (1970). The empire builders: A history of the Bolivian rubber boom and the rise of the house of Suárez. *Journal of Latin American Studies, 2*(2), 113–146.

Fundação Palmares. (2023). *Quadro geral de Comunidades Remanescentes de Quilombos (CRQS)—informações atualizadas até 31/10/2023*. Retrieved November 20, 2023 from https://www.gov.br/palmares/pt-br/departamentos/protecao-preservacao-e-articulacao/QUADROGERALPDF.pdf

IBGE, Instituto Brasileiro de Geografia e Estatística. (2017). *Censo Agropecuário 2017—Resultados definitivos*. Retrieved 20 November, 2023, from https://cidades.ibge.gov.br/brasil/ro/pesquisa/24/76693

IBGE, Instituto Brasileiro de Geografia e Estatística. (2023). *Censo Demográfico*. Tabela 1288—População nos Censos Demográficos por situação do domicílio. Retrieved November 20, 2023, from https://sidra.ibge.gov.br/tabela/1288

MapBiomas Project. (2023). *Collection 8 of the annual series of land use and land cover maps of Brazil*. Retrieved November, 20, 2023, from https://brasil.mapbiomas.org/en/estatisticas/

Neves, E. (2000). *O velho e o novo na arqueologia amazônica*. Revista USP, São Paulo, n.44, p-86-111, dez/fev 1999–2000.

Rondônia / Secretaria de Planejamento e Coordenação Geral. (1989). *Plano Agropecuário e Florestal de Rondônia - Planafloro*. Porto Velho.

Sá, LV. (2009). *Rondon: o agente público e político. Tese (Doutorado)*. Faculdade de Filosofia, Letras e Ciências Humanas. Universidade de São Paulo.

Strauss, L. (1986) [1955]. Tristes Trópicos. Lisboa/São Paulo, Ed. 70/Martins Fontes.

Vergara, M. D. R. (2013). A Exploração dos rios Amazonas e Madeira no Império Brasileiro por Franz Keller-Leuzinger: imprensa e nação. *Almanack, 6*, 81–94.

Part II
A Rethink of Cities in the Brazilian State of Rondônia: Conceptual Proposals

Chapter 5
Water-Sensitive Urban Design in Porto Velho

Anshia Singla and Özge Aktürk

Abstract This chapter focuses on the capital of the state of Rondônia, Porto Velho, in the Amazon region, to develop solutions for the water system. Through green and blue infrastructure, it aims to address urban flood adaptation measures. To this end, water-sensitive urban design solutions have been developed to meet the needs of the different zones. Due to its proximity to the Madeira River, the city faces increasing risk of flooding, as climate change increases. This chapter analyses different adaptation strategies to reduce the impact of urban flooding. The methodological framework includes literature review, mapping, and analysis through the use of remote sensing technology and design tools. The strategies include improving urban design and creating flood-resilient construction, early warning systems, sustainable drainage, and community involvement programmes. By implementing a comprehensive flood adaptation policy, Porto Velho aims to increase its resilience, protects vulnerable social groups, and promotes sustainable urban growth in the face of increasing flood risks.

Keywords Porto Velho · Urban floods · Climate change · Sustainable drainage · Floodplain management

5.1 Introduction

The Amazon River Basin is a dynamic ecosystem, and much scientific research is being done in order to understand its water cycles (Paca et al., 2020; Tian et al., 2021; Giffard et al., 2019).

It is quite evident that climate change and land use changes, such as deforestation and the expansion of agriculture and cattle ranching, are affecting the ecosystem (Figueiredo et al., 2020). The Amazon River Basin affects the hydrological cycle at

A. Singla (✉) · Ö. Aktürk
RWTH Aachen Univeristy, Aachen, Germany
e-mail: anshia.singla@rwth-aachen.de

V. G. Donoso, C. Reicher (eds.), *Rethinking Amazonian Cities in the State of Rondônia*, The Latin American Studies Book Series,
https://doi.org/10.1007/978-3-032-01817-5_5

regional and global scales, as it is considered to be the world's largest river system, covering ~40% of South America. The hydrological cycle of the rainforest affects the total global terrestrial evapotranspiration as well as the global moisture air mass fluxes (Vergopolan & Fisher, 2016; Field et al., 1998; Tian et al., 2021). Floods are also particularly relevant in this changing scenario. Research has shown that extreme droughts are becoming more frequent and intense. Urban areas are particularly affected by climate hazards such as flooding and sanitation.

The Amazon River has the largest river by volume in the world, contributing about 17% of the world's river discharge to the major oceans (Giffard et al., 2019). The Madeira River is the largest tributary of the Amazon River, representing 23% of the Amazon basin watershed. It flows northward, in a total of 3352 km from Bolivia to Brazil. It also was the largest water volume and sediment flow among its tributaries, counting about 50% of the Amazon River's sediment (Li et al., 2020).

The state of Rondônia is located in the Amazon region in the northern part of Brazil. The region of Rondônia was the land of several indigenous groups until the arrival of the colonizers. This territory was first part of other states, such as the State of Amazonas and the State of Mato Grosso. It was the result of a political agreement between Brazil and Bolivia for the construction of the Madeira-Mamoré Railroad (Petrópolis Agreement, 1903), which was also related to the history of its capital, Porto Velho. It was only in 1943 that the state region was transformed into its own territory, named Guaporé Territory. In 1956, the name was changed to Rondônia, in honour of the marshal Cândido Rondon, who explored the region (IBGE, 2023).

The capital of Rondônia state, Porto Velho, is the major city in the Madeira watershed. It has a population estimated at 519,531 inhabitants for the year 2019, and in 2000 the population was 334,661 (IBGE, 2023). It has an area of 34,096 km^2—the largest Brazilian capital in territorial size. The History of the Madeira Mamoré Railway and the rubber production are interlinked, that is why it was chosen as the site for the construction of the port, which was also a hub for transporting the rubber extracted in the forest areas to be sent to international markets, such as the USA and Europe.

The water supply in Porto Velho began with the construction of three reservoirs to supply water to the families of the workers of the Madeira-Mamoré Railway. The first reservoir was built in 1910, and the other two in 1912. The capacity of each reservoir was 200,000 L, which supplied water to the capital by gravity until 1957 (STRM Data). Two recently constructed run-of-the-river dams (Santo Antônio and Jirau), along the Madeira River in Brazil, have been controversial due to their large unquantified impacts on (1) land use and land cover (LULC) and (2) on the area that would be flooded (Emilio Moran, MDPI).

Porto Velho rainfall regime is defined by a rainy period from November to April and a dry period from June to September, with May and October considered as transitional months (Bezerra et al., 2010).

Brazil in general has a very vast geography and a very irregular development pattern, therefore urban planning has to be planned for each specific situation. The urban issues in Rondônia cities area related to the diversified and uneven distribution of resources, and it also leads to unplanned growth of the city, resulting in rapid urbanization.

In Rondônia, the specific reasons for the rapid urbanization were the enormous increase in the migrant population, which led to an increase in the need for infrastructure and employment opportunities, as well as the need to convert forest land to agricultural land and build additional houses (Moran, 1993), which were direct results of the Brazilian colonization programme implemented in Rondônia since 1964. This increased deforestation and forest fires, disturbing the ecological balance in the rainforest (REUTERS, Jake Spring). Agriculture expansion in the Brazilian Amazon has been driven not only by large-scale agriculture and cattle ranching, both of which have put considerable pressure on remaining forests and their watersheds.

5.2 Historical Background

Porto Velho has deep historical roots, dating back to the monumental construction of the Madeira-Mamoré railway in the year 1907. It was this ambitious project that set the stage for the city's development, a transformation that would shape its identity and trajectory. Although Porto Velho was not officially recognized until 1914, the seeds of its establishment as a major urban centre had already been sown (Britannica, Amy Tikkan).

The pivotal year of 1934 brought about a significant change, as Porto Velho assumed the prestigious position of being designated as the capital city of Guaporé. This strategic transition marked a turning point in the city's course, consolidating its administrative importance and putting it on the regional stage. However, it was the year 1987 that cast a long shadow over the city's environmental equilibrium. The initiation of large-scale deforestation activities triggered a domino effect that had far-reaching consequences for both the city and its surrounding ecosystem. The effects of this decision would reverberate through time, reshaping landscapes and ecological dynamics (ESA, Lars From).

The early 1990s ushered in a new wave of change for Porto Velho—this time driven by an influx of immigrants seeking opportunities and a better quality of life. This population surge put Porto Velho in the spotlight as an emerging urban centre, requiring a comprehensive reassessment of its infrastructure, services, and urban planning strategies. The city found itself at a critical juncture—how to accommodate this population boom while respecting the environment and striving for sustainable growth.

In essence, Porto Velho's journey from its origins during the monumental construction of the Madeira-Mamoré Railway to its current position as a recognized urban entity is a story of adaptation, resilience, and profound transformation. The city's historical timeline is a testament to the complex interplay between human ambitions and the delicate balance of the environment.

5.3 Research Question

Our study began with an in-depth study of the urban area of Porto Velho, with a focus on understanding the water-related issues. The analysis allowed the formulation of seven research questions, each one revealing new levels of challenges to guide possible solutions to improve the urban water problems. The research questions are:

1. What are the main needs to manage urban flooding in the case of Porto Velho?
2. Who are the main actors involved in the city of Porto Velho?
3. How can flood adaptation measures be implemented in cities?
4. What are the opportunities associated with adaptation strategies for flood management?
5. Depending on the context, which measures dealing with urban flooding can be considered the most effective to manage floods in Brazil?
6. What are the main advantages and limitations of these solutions?
7. How can these measures be applied in the spatial planning of concrete areas of the city of Porto Velho? What about the recommendations to implement these measures in Master Plans and other land policy plans?

Flood mitigation is a crucial front in the conflict over growing climate-related issues. Our study carefully examined the range of tactics used around the world and analysed how they could be modified to meet the needs of places like Porto Velho. In order to tailor adaptation strategies to the city's specific characteristics, we attempted to create a comprehensive roadmap that included everything from nature-based solutions to infrastructural fortifications.

Opportunities often emerge in the midst of difficulties, so our research delved into the realm of opportunities associated with adaptation strategy. These opportunities lie within the scope of flood management and mix with broader urban development goals. Our study sought to uncover the hidden opportunities that emerge when adaptation techniques serve as proactive, rather than simply reactive, growth engines for the whole person.

Our study began as an academic quest to sort through the maze of problems related to urban flooding in Porto Velho. Through methodical research, it was possible not only to answer these questions but also to tie them all together into a cohesive story. This story acts as a compass to guide Porto Velho towards a more flood-resilient future by highlighting the needs, stakeholders, solutions, opportunities, and constraints. More information on the methodological approach is provided below.

5.4 Process and Methods

This study utilized a combination of quantitative and qualitative research methods to gain an understanding of the research topic. The aim was to triangulate findings from sources and perspectives ensuring a rounded analysis.

Both qualitative and quantitative data were thematically analysed. The transcripts underwent coding using both deductive approaches. Through a process of data immersion, categorization and constant comparison, themes, and patterns were identified.

The methodology employed analysis and mapping tools. The former played a role in understanding the city, its region, history, local legislation, and current challenges. Additionally bibliographic research helped uncover indicators related to sanitation in Porto Velho as well as economic factors such as the Gross Domestic Product (GDP) of the municipality for a specific year. With this information at hand, the following steps were undertaken;

Analysis of how basic sanitation rankings corresponded to indicators.

Description of issues within water supply, sewerage systems, urban drainage, and solid waste management.

Projection analysis based on required investments, for universalizing services alongside current investment trends.

To further explore and analyse the city and its surroundings we utilized mapping as a tool. We incorporated layers of data, including water bodies, green spaces, topography and transportation infrastructure, such as roads and railways using McHarg (1969) landscape analysis technique.

Spatial syntax and morphological analysis have also been applied in order to provide a comprehensive approach to urban reality. This methodological journey helped in concluding the solutions for the drainage problems in the city, for that there was synergizing multiple layers of understanding into a cohesive strategy.

Thereafter, space syntax method was adopted, a method that considers the analysis of the cityscape. Paths, corridors, "axial lines," "convex spaces" are considered in this approach. These axial lines, like arteries of movement, reveal the primary routes that dictate the flow of the city. Meanwhile, the "convex spaces" offer a glimpse into the central hubs and places where the convergence of axial lines creates a nexus of activity. This perceptive analysis forms the basis of our subsequent strategies. The morphology analysis method was used for the understanding of the city's past and present urban form, urban structure, land use, and patterns.

Some of the analyses resulting from the research methodology are detailed below.

5.5 General Analysis

5.5.1 *Topography*

Porto Velho's topography is predominantly flat, with gentle slopes and occasional low hills. The topography is largely shaped by the presence of rivers, streams, and wetlands, which contribute to the region's rich biodiversity and ecological importance. However, the flat terrain and the city's location in a flood-prone area also pose challenges for urban planning and infrastructure development, particularly in mitigating the risks associated with annual flooding events. Inundation hazards are caused by seasonal rainfall and river flooding. Levees, drainage systems, and urban design are among methods used to mitigate the effects of flooding. Resilience strategies are critical to safeguard the city from disasters.

5.5.2 *Water and Dam*

The city of Porto Velho is surrounded by the Amazon River and Madeira River. Important Dams in Porto Velho are Jirau and Santo Antonio. These are the two important structures which help in flood control in the city and also acts as a major source of transport and economic activities. Furthermore, these are the two major barriers which help in flood control. This important body of water supports a variety of habitats in addition to facilitating transit. However, these two dams have a huge impact on the biodiversity and the social groups, there have been also serious threats and flooding due to the construction of these two dams, which resulted into social and environmental problems. Furthermore, the energy produced through these dams does not satay in the region and is being sold out, which is why the regions suffers from energy instability. See (Fig. 5.1) for the placement of the Dams and the river flowing through the region.

5.5.3 *Urban Growth*

Socio spatial formation of the city of Porto Velho, the period between 1980 and 2010 played a big role in realizing the Urban Reality of Rondônia, Carlos Santos, Maurício Silva Published 29 June 2012). As we see in Fig. 5.2. Over the years people started moving on the outskirts of the city in search of better opportunities, which created an Urban arch where there are clusters of settlements of people. The municipality has experienced cycles of deforestation and the construction of projects, roads and dams, which have altered social conditions and land use. Since 2000, Porto Velho has undergone a particularly dramatic transformation due to the construction of two hydroelectric power stations (HPS) along the Madeira River.

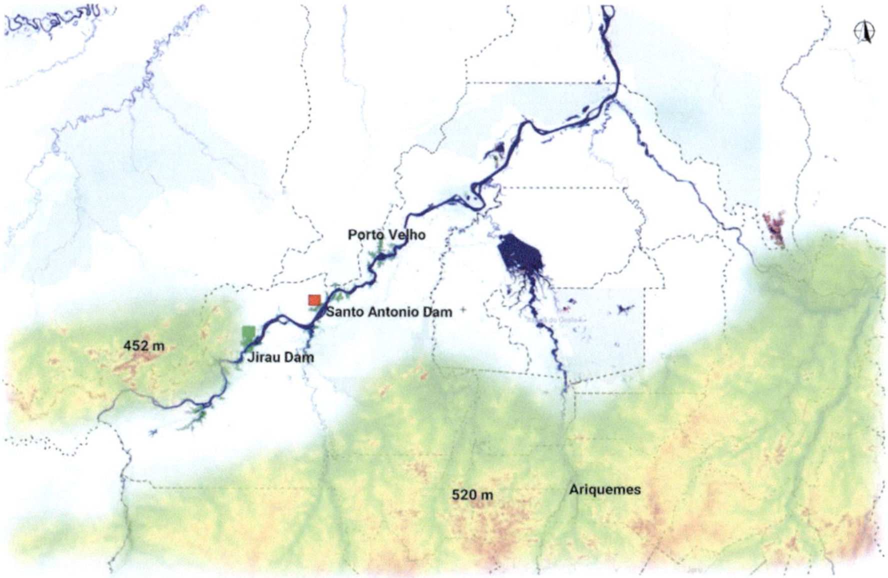

Fig. 5.1 Porto Velho's Water body and dams. (Source: Elaborated by Singla and Aktürk, based on data from Mapbiomas (2021) and IGBE)

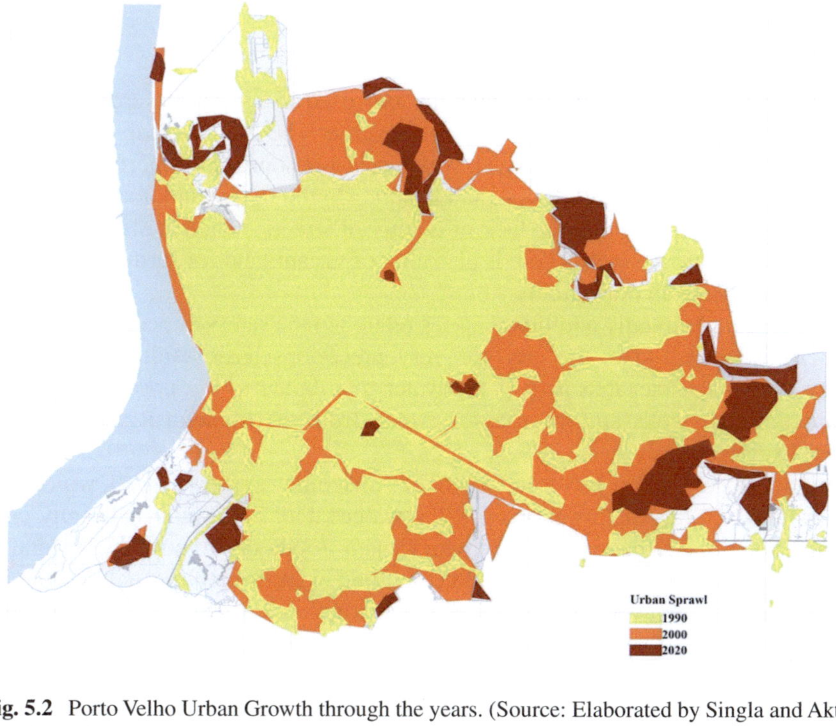

Fig. 5.2 Porto Velho Urban Growth through the years. (Source: Elaborated by Singla and Aktürk, based on information from the local administrative site of Porto Velho municipality, Rondônia State)

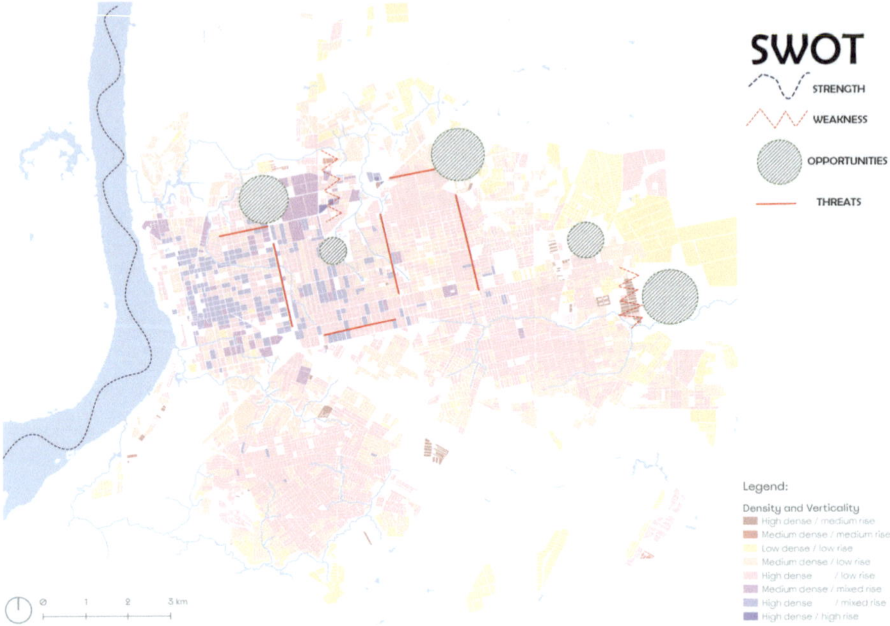

Fig. 5.3 Swot Analysis of Porto Velho. (Source: Elaborated by Singla and Aktürk with the support of Denis Eke)

5.5.4 SWOT and Challenges

A SWOT analysis is a strategic planning tool used to assess the area strengths, weaknesses, opportunities, and threats of an area. The SWOT analysis of Porto Velho showed the strength of the river, which acts as a source of important activities for trade, commerce, transportation, and leisure. On the other hand, the weakness is the poor urban planning and the lack of quality of streets, which leads to road congestion and drainage issues. There is also a lot of vacant land for further expansion, which is definitely an opportunity.

There are undoubtedly additional issues related to the safety of people, due to the high crime index.[1] Also, the city has very precarious basic sanitation services, aggravated by the fact that part of the water distributed to the population comes from deep tubes wells and, to a lesser extent, from the Madeira River (IJEARS, Kleber Borges).

Part of the population gets water from the Amazonian system as well, which collects water from the shallowest part of the water table. This water is usually contaminated and the situation is exacerbated when floods occur, such as the one in 2014. Irregular urban occupation and difficulties in implementing public policies make the population even more vulnerable to urban flooding (WWF Report, Kricher 1997). See Fig. 5.3 for a clearer understanding of the SWOT analysis.

[1] The homicide rate in Porto Velho was 27.38 for the year 2019. For comparison, São Paulo had a rate of 3.71 for the same year (IPEA, 2019).

5.6 Plan of Action

In the initial phase of our project, we conducted a morphological analysis that included the classification of various building typologies and their surroundings. This revealed a notable prevalence of high-rise and mixed-use buildings in the urban landscape. This observation also highlighted the emerging pattern of clustering, which contributes to the creation of open spaces. Unfortunately, these open spaces have yet to reach their full potential and remain underutilized. This presents an exciting opportunity for innovative interventions that could revitalize these spaces and transform them into vibrant community spaces.

Our focus then shifted to a comprehensive review of the flood risk plan for the area. This strategic approach provided valuable insights into the different flood-prone zones in the region, allowing the formulation of strategies to improve the resilience and preparedness of these specific zones in anticipation of potential flood events.

The flood risk analysis revealed a number of challenges and vulnerabilities that required immediate attention. Identifying these critical issues allowed us to determine the precise actions needed to effectively strengthen flood mitigation and adaptation efforts. With an informed approach, we set out to design tailored solutions that not only address existing vulnerabilities, but also lay the foundation for a more sustainable and secure urban environment.

We then transitioned to the creation of a comprehensive Vision Plan, meticulously tailored to the unique needs of each designated zone. This plan embodied our commitment to holistic urban development and served as the blueprint for a systematic implementation strategy, integrating existing infrastructure, land use patterns, and drainage systems.

In summary, our exploration with the categorization of building typologies and spatial configurations, revealing the trend of clustering and latent open spaces. This discovery prompted a closer examination of the flood risk plan, decoding the specific flood-prone zones and their immediate fortification needs. The next step was the creation of the Vision Plan, positioning us at the threshold of a transformative journey towards urban improvement.

Some water-sensitive regions were selected as implementation zones in the structure plan we developed for the city centre of Porto Velho. Zones A, B, C, D, and E were recognized in Fig. 5.4 as being significant based on several criteria. Some factors considered for selecting these areas included prior floodplains, infrastructure, and closeness to the city core. In the structural plan, three different street kinds were chosen: low density A type streets, which are coloured blue, medium density B type streets, which are coloured green, and high-density C type streets, which are coloured red (Fig. 5.4).

In the plan, the igarape (waterways) in the region was another determining factor in the zone selection. The distance or closeness of the igarape to high-density areas was taken into consideration as it would affect the sustainable solutions planned.

Various evaluations were done, including distribution of public spaces, utilization characteristics of the regions, such as whether they are used for residential or industrial purposes, and green-oriented zones within the zones.

Another feature that distinguishes the selected regions from the others is that they have different requirements and are important against the risk of flooding. In the upcoming years, other regions will join these ones in contributing to the designs for the new master plans. Regions that have different needs and are guiding are chosen as examples for future designs.

For instance, zone C is situated near the Madeira River and features a moderately dense green area. This area contrasts with zone A, which has more intense urbanization and a drain that runs through it. Although these two places have very different demands, the flood danger is about the same. The specifics of our studies have gone in this direction, as you can see. Details of these regions is found in the following section (Implementation zones).

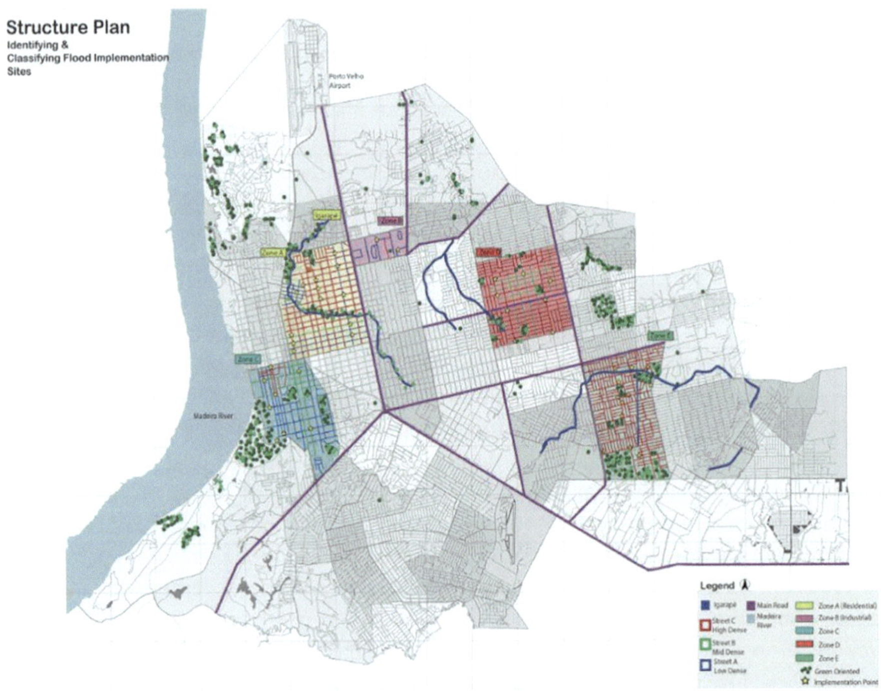

Fig. 5.4 Structural Plan: identifying and classifying flood implementation areas. (Source: Elaborated by Singla and Aktürk. Created using QGIS 3.28.0 version)

5.7 Implementation Zones

In our comprehensive implementation strategy, we have defined different typologies for each zone, taking into account various factors that influence their functionality and vulnerability to flooding. Using a standardized scale ranging from 1 to 10, the vulnerability to flooding was assessed for each zone, allowing for prioritization of zones.

In addition, a density analysis was conducted for each zone to assess the concentration of infrastructure, buildings, and population within the boundaries of the zone. This density assessment serves as a critical determinant in the design of solutions, guiding the allocation of resources and the implementation of measures that address the unique needs of each zone.

The classification of the zones was based on their respective land uses. This segmentation ensures that the solutions are tailored to the specific functions and activities that take place within each zone, optimizing the impact of the interventions.

In addition, the street selection process aimed to identify key thoroughfares within each zone where innovative solutions could be applied. These streets serve as vital conduits for connectivity and accessibility, and by focusing our efforts here, we aim to create a ripple effect that extends the benefits of our interventions throughout the entire zone. For a more detailed representation of the characteristics and intricacies of each zone, see Fig. 5.5. This illustrative representation reunites the essence of our zoning framework and visually conveys the nuanced attributes that have informed our implementation strategy.

If we examine the specifics of the chosen zones, they are:

Zone A (Sao Joao Bosco): This zone includes the big igarape and high-rise buildings. The big residential and commercial area can cause high risk of damage due to flood. Flood risk in this area is 8/10. Also the Flood in 2014, damaged this area. Density in this area according to the google maps site view is 7 out of 10. Defining this zones type is residential and commercial. Most of the governmental buildings and the economic activities are going in this zone and this cause a big movement of people and density.

Zone B (R. Cipriano Gurgel): This is an industrial zone of our structure plan. Choosing this area may lead us the different needs of an industrial zones. In industrial zones, transportation issues and density are different from other zones. Flood risk in this area is 8/10. Flood risk will affect the low-rise depot stores in this zone. In order to prevent new problems in industrial zones specified sustainable drainage systems are needed. Density in this area according to the google maps site view is 5 out of 10.

Zone C (Santa Barbara): This zone is the closest to the Madeira River. It includes green zones and it has dense residential area. Flood risk in this area is 9/10. According to the analaysis locating the end of the igarapes or closest to the water areas may cause higher damage for residential buildings. Density in this area according to the google maps site view is 8 out of 10.

Fig. 5.5 Implementation zones and classifications. (Source: Elaborated by Singla and Aktürk, based on Google maps)

Zone D (Esperança da Comunidade): This zone includes highest populated neighbourhood area. High dense neighbourhoods are important to consider as daily life flood problems such as backyard water drainages, car park drainage areas or walking paths. Density in this area according to the google maps site view is 9 out of 10. Flood risk in this area is 7/10. Brazilian neighbourhood designs are the key factor in the solution making.

Zone E (Tancredo Reves): This zone has the second big igarapé and huge public spaces. Density in this area according to the google maps site view is 8 out of 10. Public spaces main needs should be defined by residents and decision makers in the zone. Flood risk in this area is 7/10.

To sum up, there is a correlation between zone density and a higher risk of flooding. In conclusion, Porto Velho's classification of these five zones offers important insights into the different degrees of flood risk and urban features found in each area. It is clear that each zone requires unique techniques, from targeted urban design concerns in densely populated neighbourhoods to sustainable drainage systems in industrial areas. Urban planners and authorities can develop specialized solutions to improve flood resilience and protect the well-being of Porto Velho's residents by knowing the distinctive characteristics and dangers associated with these zones.

5.8 Conclusion

Addressing urban flooding in Porto Velho requires a multi-faceted approach that considers the city's unique needs and challenges. To conclude, in order to manage the urban floods, preparing a new structural plan according to the city's needs is important and also analysing the flood risk areas according to different criterias plays a vital role.

The main needs for managing urban flooding in Porto Velho include improving drainage systems, enhancing flood monitoring and early warning systems, and implementing sustainable land use planning. Preparing a new structural plan with using zoning framework will help the water-sensitive issues. Design positive structural plan will lead us to defining problematic areas quickly and easy to implement solutions.

Local government representatives, urban planners, civil engineers, environmental agencies, and community people are among the essential participants in this project. These parties must work closely together to create and put into action efficient flood adaptation strategies. Infrastructure improvements, land use restrictions, and community involvement programmes can all be used to carry out these measures. Increased resilience, better urban planning, and potential financial gains are advantages linked with adaptation measures for flood management. The development of green infrastructure, the adoption of construction rules that are flood resistant, and zoning laws that protect floodplains could represent the most successful strategies in the Brazilian setting. These solutions have benefits like increased public health and decreased property damage but they also have drawbacks like high costs and difficult implementation.

Acknowledgements The authors thank (May Masuda and Denis Eke), students of the master programme "Transforming City Regions" (RWTH Aachen University, Germany) for their contribution with the morphological analysis of Porto Velho.

References

Field, C. B., et al. (1998). Primary production of the biosphere: Integrating terrestrial and oceanic components. *Science, 281*(5374), 237–240. https://doi.org/10.1126/science.281.5374.237

Figueiredo, R. D. O., Cak, A., & Markewitz, D. (2020). Agricultural impacts on hydro biogeochemical cycling in the Amazon: Is there any solution? *Water, 12*, 763. https://doi.org/10.3390/w12030763

Giffard, P., Llovel, W., Jouanno, J., Morvan, G., & Decharme, B. (2019). Contribution of the Amazon River discharge to Regional Sea level in the tropical Atlantic Ocean. *Water, 11*, 2348. https://doi.org/10.3390/w11112348

IBGE. (2023). *Brazilian Institute of Geography and Statistics*. Retrieved September 1, 2023

IPEA. (2019). *Atlas da violência*. Retrieved September 1, 2023, from https://www.ipea.gov.br/atlasviolencia/dados-series/20

Li, D., Lu, D., Moran, E., & da Silva, R. F. B. (2020). Examining water area changes accompanying dam construction in the Madeira River in the Brazilian Amazon. *Water, 12*, 1921. https://doi.org/10.3390/w12071921

McHarg, I. L. (1969). *Design with nature*. Doubleday/Natural History Press.

Paca, V. H. D. M., Espinoza-Dávalos, G. E., Moreira, D. M., & Comair, G. (2020). Variability of trends in precipitation across the Amazon River basin determined from the CHIRPS precipitation product and from station records. *Water, 12*, 1244. https://doi.org/10.3390/w12051244

Tian, K., Wang, Z., Li, F., Gao, Y., Xiao, Y., & Liu, C. (2021). Drought events over the Amazon River basin (1993–2019) as detected by the climate-driven Total water storage change. *Remote Sensing, 13*, 1124. https://doi.org/10.3390/rs13061124

Vergopolan, N., & Fisher, J. B. (2016). The impact of deforestation on the hydrological cycle in Amazonia as observed from remote sensing. *International Journal of Remote Sensing, 37*, 5412–5430. https://doi.org/10.1080/01431161.2016.1232874

Chapter 6
Rethinking Amazon Cities from Rondônia State: The Case of Ji-Paraná

Ibtsam Saeed, Muhammad Tahir Ali, and Berfin Gül

Abstract This book chapter examines the environmental and social challenges faced by Ji-Paraná, the second most populous municipality in the Brazilian state of Rondônia. Focusing on deforestation, biodiversity loss, frequent forest fires, land conflicts, and socioeconomic disparities, this chapter analyses their causes and effects. Geospatial data and analysis techniques, including QGIS, NDVI analysis, and MapBiomas data collection, were used as a method to map and analyse the impacts of land use changes and landscape patterns. To mitigate the challenges identified, this chapter proposes urban planning tools to address urbanization-related issues, combined with biodiversity strategies that include the protection of natural areas, the promotion of sustainable agriculture, reforestation, and sustainable urban planning. Key areas include comprehensive urban planning, efficient land use, provision of green spaces, and improved transport infrastructure. Social strategies such as the creation of cultural hubs and community engagement are also highlighted as tools to promote inclusiveness and effective conservation efforts. By integrating scientific research, geospatial analysis, and community engagement ideas, this chapter offers practical approaches to support sustainable development and conservation efforts in the region. By using mapping techniques, data analysis and participatory methods, decision-makers can make informed choices for a resilient and equitable future in Ji-Paraná.

Keywords Amazon · Rondônia · Ji-Paraná · Sustainable development · Urban design

I. Saeed · M. T. Ali (✉) · B. Gül
RWTH Aachen University, Aachen, Germany
e-mail: ibtsam.saeed@rwth-aachen.de; muhammad.tahir.ali@rwth-aachen.de

V. G. Donoso, C. Reicher (eds.), *Rethinking Amazonian Cities in the State of Rondônia*, The Latin American Studies Book Series,
https://doi.org/10.1007/978-3-032-01817-5_6

6.1 Introduction

The Amazon rainforest has experienced widespread loss of resilience since the early 2000s, particularly in drier areas and regions affected by human activity (Boulton et al., 2022). Within this vast region, Rondônia, in western Brazil, is one of the most heavily deforested states (Earth Observatory.nasa.gov, 2023; Forest champions, 2020; SPA, 2021), where the expansion of human settlements has caused significant changes in forest cover patterns (Jacques Imbernon, n.d.). Natural forests are being cut down, and an expansion of the agricultural lands can be observed (Fearnside, 1984; Fearnside & Cruz, 2018). Ji-Paraná, located in the east-central portion of the state of Rondônia, is an example of this change. It is a dynamic city that has had remarkable growth in recent years. However, this progression has been accompanied by its own set of challenges. Deforestation associated with inappropriate land use patterns are potentially dangerous for the climate and could disrupt the current ecosystem balance.

The analysis of the urban form of Ji-Paraná's reveals the scarcity of green spaces in the city as well as the lack of accessible social and public spaces. This not only results in a lack of space for social interaction and recreation but it also highlights the importance of community participation in the development of the city. In addition, deforestation plays a critical role in the larger scale of the municipal area.

Mapbiomas (2021) data imagery allows us to see the timeframe of the land use changes: notably, in the period from 2002 to 2022, the loss of primary forest in Ji-Paraná increased from 3.8 hectares to 900 hectares. In addition, the municipality suffered a loss of 2.56 hectares of tree cover due to fires and 31.6 hectares due to other factors between 2001 and 2022.

Confronting the challenges of Ji-Paraná and catalysing its transformation into a more desirable living space for its inhabitants is the essence of this book. At the core of our work are urban design and sociocultural proposals that can create transformative change. Central to this mission is the creation of inclusive public spaces that are universally accessible and reflect the diversity of the community. In this quest, the importance of incorporating local perspectives, views, and active participation is vital.

We cannot ignore the pressing issue of deforestation, which poses a serious threat to the region's delicate ecological balance. It is crucial that we take immediate action to protect and restore the natural resources that define the biodiversity and uniqueness of Ji-Paraná. By doing so, we can support the sustainability and resilience of society. Throughout our research, we highlight the tangible benefits (social, cultural, and economic) that result from the creation of these inclusive public spaces. Our mission to improve the urban fabric of Ji-Paraná is linked to a sustainable coexistence, fostering an environment where both society and nature can thrive in harmony.

6.2 Materials and Methods

The first methodological approach of this book chapter was structured around a set of guiding research questions (Fig. 6.1). The method was divided into two parts. The first part focused on researching the region, guided by questions such as: "What are the main causes of the problems?" After collecting information about the existing conditions and identifying the key challenges of the region, the second part addressed these issues by developing strategies for the sustainable development of Ji-Paraná. This stage was guided by questions such as: "How do we want to solve these problems?"

With some answers to the questions, a mixed method approach in four stages of research (Fig. 6.2) was combined in two scales of action, regional (state of Rondônia), and the city (Ji-Paraná).

On stage 1, literature review and bibliographic research methods were used to collect quantitative and qualitative data. On stage 2, data collection in geographic information from satellite imagery was gathered from several sources, such as Mapbiomas (2021), ANA (2023), USGS Earth explorer (2023), Esri: living atlas (2017), EU Copernicus (2023) were combined together to allow spatial data visualization. In the third step, the QGIS software, Google Earth Engine, Adobe photoshop, and Adobe illustrator were used to visualize and analyse the data using

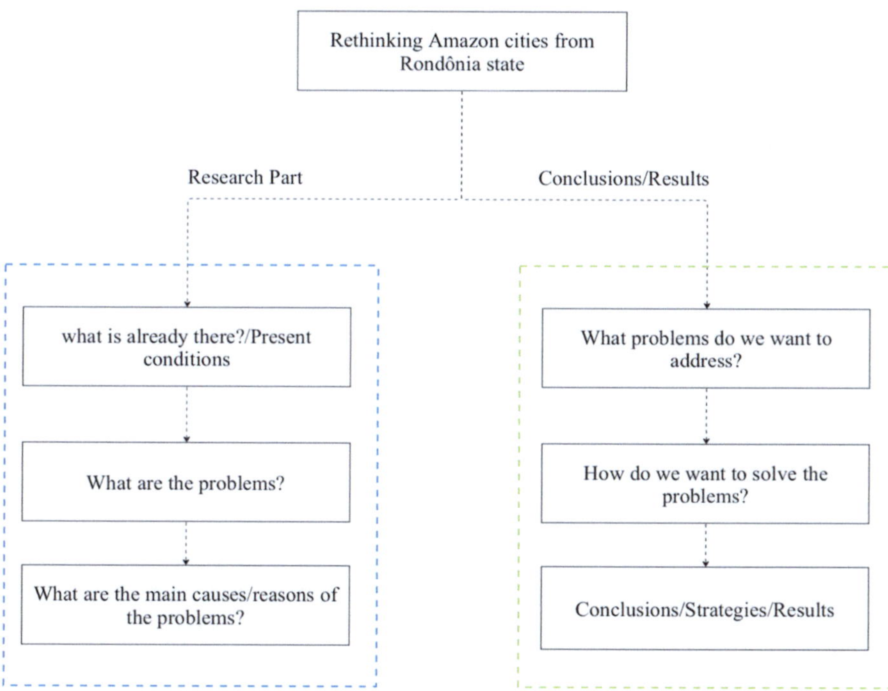

Fig. 6.1 General approach. (Source: Elaborated by Saeed, Tahir Ali, and Gül)

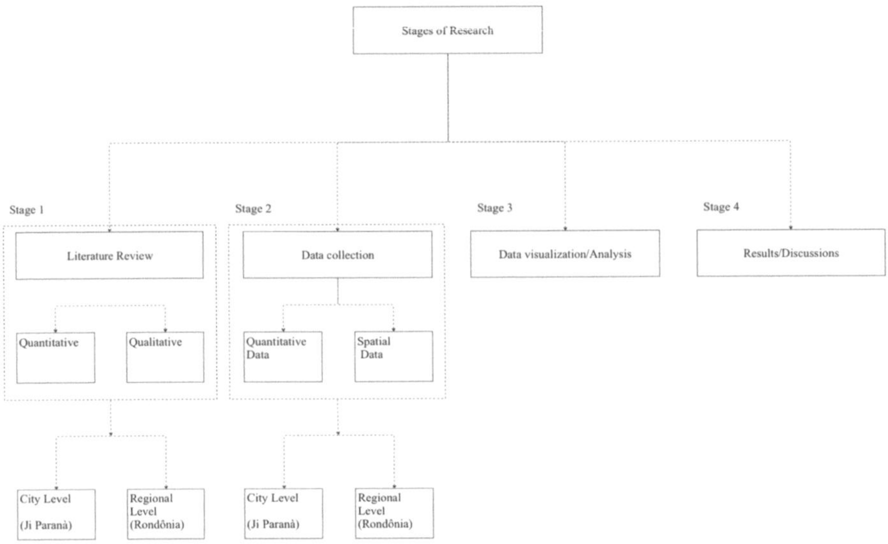

Fig. 6.2 Stages of research. (Source: Elaborated by Saeed, Tahir Ali, and Gül)

different tools and techniques. Stage 4 consisted in discussing the data and proposing solutions at the urban scale.

The detailed information of each step is presented below:

6.2.1 Stage 1: Literature Review

The first stage—literature review—was carried out by selecting scientific literature (articles, reviews, documentaries, and others) from various online data sources, such as the Research Gate platform, Global Forest Watch (2023), Mapbiomas (2021), and the Brazilian Institute for Geography and Statistics (IBGE, 2023) and Youtube documentaries. Some specific keywords, such as "Rondônia State", "Deforestation", "Amazon", "Rondônia: Land use", "Rondônia: Main issues", "Ji-Paraná history", "Ji-Paraná: Land use", and "Ji-Paraná issues" were used to find scientific publications and other resource materials. Using these keywords, we were able to identify the main issues/main problems, both in terms of quantitative (number-based) and qualitative (textual and descriptive) data, mainly at the regional level, but negligible at the city level. For example, one of the main issues we found at the regional level was "deforestation". The other issues, such as environmental and social issues, were also related to deforestation.

Further literature review using other keywords gave more insights about regional problems. For example, "Loss of biodiversity", "flooding and droughts", "forest fires", and "pollution".

6.2.2 Stage 2: Data Collection

It was also important to analyse how the deforestation, environmental, and social issues we identified at the regional level were linked at the city level. Nevertheless, most of the data which was collected through the literature review was only available at the regional/ state level, so it was important to also collect data at the city level as well.

Therefore, we collected more quantitative spatial data from satellite imagery at open-source platforms such as the MapBiomas (2021), United States Geological Survey: Earth Explorer (USGS, 2023), Esri: Living Atlas (2017), and Copernicus EU (2023).

For quantitative data regarding deforestation, we used MapBiomas (2021) and Global Forest Watch (2023) websites to estimate the forest loss since 1985 till present period, and also the percentage of remaining forest areas. Other data collection was also gathered, to measure precipitation and land surface temperature, at both regional and city scale.

For Spatial data regarding the land use and land use change, we used the satellite imagery which was collected from the USGS: Earth explorer, Esri, Living Atlas, and other Copernicus websites. The satellite imagery was collected using two different satellite systems: Landsat 7 and Sentinel 2. It is also important to note that the Sentinel imagery was archived from the USGS website in October 2022 and therefore it is only available on Copernicus and Esri: living atlas websites.

6.2.2.1 Data Characteristics

Landsat 7 satellite imagery was collected from the USGS: Earth explorer website, Landsat collection 2 Level-2 (Landsat 7 ETM+ C2L2) with the spatial resolution of 30 m x 30 m and with cloud coverage between the range of 0% and 30%. The Sentinel imagery was collected from Esri: Living atlas (Sentinel 2, 10 m Landuse and Land cover tiles) The sentinel imagery on the Esri website is only available from the year 2017–2022 and in order to get the imagery of the previous years we used the Copernicus website. Apart from this the forest loss and tree cover datasets from the website global forest watch is also from Landsat satellite images with 30 m x 30 m resolution.

6.2.3 Stage 3: Data Visualization/Analysis

To visualize the data collected from the satellite imagery and other open-source platforms, we used the open-source software QGIS, along with Google Earth Engine to handle the open data. In order to enhance the created graphics, we used Adobe Photoshop and Illustrator softwares. In addition, some plugins were also used during the visualization process like QuickOSM, Active fire, and SRTM downloader. The data visualization/analysis was further divided into different steps to understand the region and the city. The following steps were performed using the satellite data during the third stage.

6.2.3.1 Digital Elevation Model (DEM)

A Digital Elevation Model (DEM) was created in QGIS using the satellite data to visualize the topographic surface at the regional and city level. The produced DEM gave the information about the topography pattern for Rondônia and Ji-Paraná.

6.2.3.2 NDVI Using Remote Sensing in Qgis

The Normalized Difference Vegetation Index (NDVI) uses near-infrared and red light to quantify and measure the vegetation (GISGeography, 2023). NDVI is a standardized way to measure healthy vegetation. High NDVI values demonstrate healthier vegetation, and lower values demonstrate less or no vegetation. NDVI calculations were performed using Landsat imagery to identify the low and healthy vegetation around the city of Ji-Paraná.

6.2.3.3 Land Use and Land Cover Change

To show how land use and land cover have changed from 1985 to 2021 at both the regional and city levels, Sentinel 2 data from Esri: Living Atlas for the years (2021) was collected and visualized. For the previous years, data from Mapbiomas (2021) was used. In this stage it was possible to calculate the percentage of the forest that was present in 1985 and how much it was reduced over the time until 2021. Deforestation was also mapped using the same process both at the regional and city levels.

6.2.3.4 Precipitation and Land Surface Temperature Visualization

In this step, the average annual precipitation was visualized on different levels and on a higher regional scale due to data availability. First, the analysis of the country's rainfall was carried out; second, the state of Rondônia and Ji-Paraná was analysed. For this visualization we used the annual precipitation dataset PERSIANN-Cloud Classification System (PERSIANN-CCS) with the global resolution of 4 km x 4 km from the centre of Hydrometeorology and remote sensing (CHRS) from different years (2005, 2015, and 2021) and afterwards the datasets were visualized in QGIS.

For the land surface temperature (LST) visualization, the platform of Google Earth Engine was used to read the satellite data. The data was visualized using the code editor of GEE and the produced results gave information about the hottest and the coldest months and the average temperature of Brazil.

6.2.3.5 Identification of Flood Prone Areas

The flood prone areas were identified using the Digital Elevation Model of Rondônia in QGIS. This data allowed the analysis of the flood risk to identify the areas which are more prone to flooding.

6.2.4 Stage 4: Results/Discussions

After completing the visualization and analysis stage, in the final stage, we discussed all the collected information based on the results which were generated using the satellite data, open data, and literature review to develop the best possible strategies for the city of Ji-Paraná. The strategies were categorized according to social, urban, environmental, and economic levels. These strategies can also be implemented at the regional level and in other parts of Brazil.

6.3 Analysis

6.3.1 History in a Nutshell

The history of Amazon's occupation began with Asian immigrants 14,000 years ago, leading to the emergence of complex indigenous societies (Neves, 2022). European colonization brought diseases and conflict, causing a decline in the indigenous population. In Rondônia, a state within the Amazon, a Brazil-Bolivia agreement in 1903 led to the construction of the Madeira-Mamoré Railroad, triggering growth and the foundation of Porto Velho as a major city.

Rondônia's history is shaped by economic cycles, including gold, rubber, and minerals. World War II triggered migration and mineral discoveries, accelerating population, and economic growth. The 1964 military coup in Brazil expanded the Amazon frontier, supported by the Transamazonica highway, leading to new settlements and deforestation. The Institute of Colonization and Agrarian Reform (INCRA) and private firms distributed land along Highway BR 364, fueling city development. This process displaced indigenous, riverside, and quilombola communities (da Cunha & Moser, 2010).

The transformation of the Amazon region is a result of economic interests, infrastructure projects, and policies. The complex interplay between economic growth, social displacement, and environmental impacts continues to shape the region's current challenges and future prospects.

6.3.2 *The Impact of Deforestation on Biodiversity*

The biodiversity of terrestrial and freshwater ecosystems in the tropical Andes and lowland Amazon is facing an alarming threat due to various human-induced factors. Deforestation, habitat fragmentation, overexploitation, pollution, and climate change are pushing many plants and animals, including mammals, birds, reptiles, amphibians, fish, and terrestrial invertebrates, to the brink of extinction. Some of these species are classified as Critically Endangered, and even endemic birds and mammals in the eastern Amazon have dwindling populations, often in the tens or hundreds. Shockingly, among the vast array of Amazonian tree species, an estimated 36% to 57% are at risk of meeting the criteria for global threat classification according to the International Union for Conservation of Nature (IUCN). It is important to note that these numbers likely underestimate the true extent of the crisis, as many Amazonian species remain unstudied and undocumented. While there's no record of regional extinctions yet, local extinctions are increasingly common, with several species confined to tiny fragments of their once-expansive habitats. Furthermore, Andean species, with their limited distributions, face a particularly high risk of extinction, exemplified by the plight of Colombian birds. The ramifications of these extinctions are far-reaching, affecting pollination and consumption networks, leading to profound shifts in ecological interactions that imperil the very fabric of Amazonian biodiversity (Adapted from the Science Planet BKG 12.1 Report 2021).

In 2020, WWF released a report highlighting the alarming consequences of the planet's biodiversity crisis, which includes the disappearance of natural habitats and the accelerated loss of species. This assessment examined different aspects of land use change, encompassing the overall impact up to 2020, the historical impact until 2014, and the recent impact from 2014 to 2020. The primary focus was on two key drivers of habitat change: pasture and soy activities.

The transformation of the Amazon's original vegetation into pastures and soybean plantations has been notably extensive. To date, approximately 17% of the forest cover has been lost, with an additional 13,235 km^2 cleared between August 2020 and July 2021 due to these activities (WWF, 2021). This shift in land use has had profound consequences for the region's wildlife. When examining the overlap between species' habitats and converted areas, a troubling pattern emerged. Nearly all of the assessed species (484 out of 486) experienced adverse effects due to deforestation, resulting in significant reductions in their habitat. Iconic species like the Brazilian Merganser, Cock-tailed Tyrant, Goias Parakeet, red-tailed lizard, and marsh deer have faced habitat losses of up to 70%.

The impact of deforestation in the Amazon extends beyond its landscapes; it directly impacts numerous species, including those that are unique to the region, act as environmental indicators, and even iconic flagship species.

The assessment included 88 endemic species within the Amazon biome, covering more than 95% of their known habitats. Worryingly, these species have suffered an average habitat loss of 17%. Among them, the Northwestern Tocantins Oval Frog

(*Elachistocleis carvalhoi*) has already lost over 80% of its habitat. Even the Brasilia Tapaculo (*Scytalopus novacapitalis*), named after its proximity to the nation's capital, has seen a significant reduction of over 56% of its original range. Notably, flagship species such as the giant armadillo (*Priodontes maximus*) and the maned wolf have also been affected by habitat loss, with over half of their habitats disappearing. The Southern Tiger Cat (*Leopardus guttulus*) and Azara's Capuchin (*Sapajus cay*) have faced nearly 80% loss, while the Southern Three-banded Armadillo (*Tolypeutes matacus*) has experienced significant declines due to habitat loss and hunting pressures (WWF, 2021).

This widespread decline in various species underscores the precarious state of ecosystems and their intricate connections. As habitats shrink, the functions and services provided by these species diminish, affecting water resources and socio-economic activities. For instance, the Brazilian Gracile Opossum (*Gracilinanus microtarsus*), despite being insect predators, confronts habitat loss, which impacts pest control dynamics in regions with soy plantations. These findings emphasize the urgent need for comprehensive conservation strategies to protect the remarkable biodiversity of the Amazon.

The cumulative impact on species equates to approximately 56.3 million km^2 of lost habitat in the Amazon by 2021. The rate of habitat loss accelerated between 2014 and 2021, resulting in an overall reduction of 2.2% of the Amazon's total area. This escalation was particularly notable in specific regions, significantly affecting certain species such as the Rondon's Marmoset. Even outside concentrated deforestation zones, species like the Lesser Nothura have demonstrated remarkable resilience within protected areas like National Parks, although they have already lost over 72% of their original habitat (WWF, 2021).

These findings underscore the urgency of conservation efforts for the unique and endangered biodiversity of the Amazon. As deforestation continues, it becomes crucial to prioritize protected areas and implement zero deforestation policies to prevent further species loss. The insights from this study emphasize the pivotal role of proactive land use planning and the safeguarding of existing converted areas to prevent the disappearance of invaluable species within the Amazon Forest (WWF, 2021).

6.3.3 Deforestation in Rondônia

Deforestation in Rondônia state has been a critical environmental issue, with devastating consequences. Since the 1970s, Rondônia has experienced alarming deforestation rates, as illustrated by Figs. 6.3 and 6.4, which show the decline in forest cover from 88.06% in 1985 to 58.20% in 2021. The main drivers of deforestation in Rondônia are agricultural expansion, logging, and infrastructure development. The lure of economic opportunities, particularly cattle ranching and soybean cultivation, has fueled rampant land clearing. Road construction, such as the BR-364 highway, has facilitated access to previously untouched areas, accelerating deforestation.

Landuse change of Rondonia State

Classe	1985 Total (Mn Ha)	Area in Percentage	2005 Total (Mn Ha)	Area in Percentage	2021 Total (Mn Ha)	Area in Percentage
Forest	20,927,738	88.06%	15,685,577	66%	13,829,956	58.20%
Natural Non-Forest Formation	796,356	3.35%	786,234	3.31%	827,277	3.48%
Agricultural	1,783,253	7.50%	7,033,961	29.68%	8,817,828	37.10%
Non Vegetable area	39,189	0.17%	54,300	0.23%	65,612	0.28%
Body of water	217,945	0.92%	204,407	0.86%	223,808	0.94%
Urbanized Area	19,423		39,186		49,543	

Fig. 6.3 Land use and land cover change (1985–2021). (Source: Elaborated by Saeed, Tahir Ali, and Gül, using data from Mapbiomas (2023))

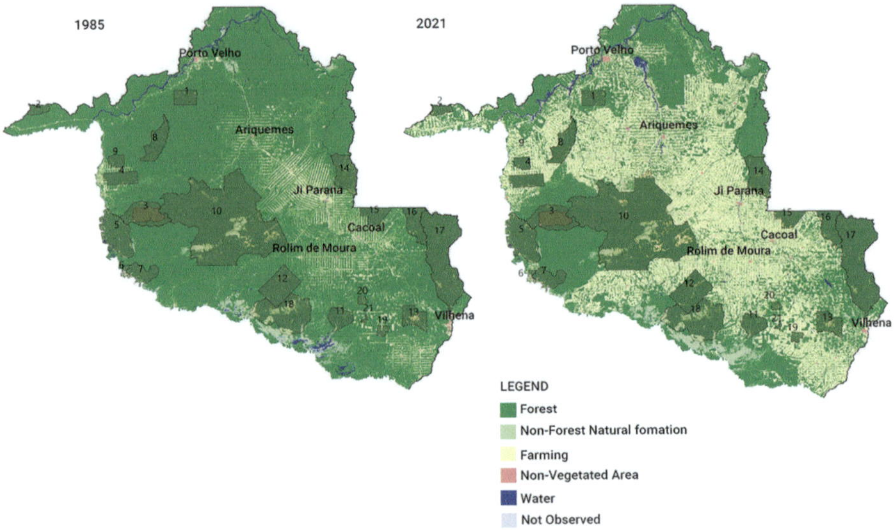

Fig. 6.4 Land use and land cover change (1985–2021). (Source: Elaborated by Saeed, Tahir Ali, and Gül, using data from Mapbiomas (2023))

6.3.4 Deforestation in Ji-Paraná

Over the years, Ji-Paraná has experienced a worrying deforestation trend. Between 2000 and 2020, the area has lost approximately 2500 square kilometres of forest cover, according to data from the Brazilian National Institute for Space Research (INPE). This alarming rate of deforestation is mainly due to agricultural expansion, illegal logging, and road construction. The main causes of deforestation in Ji-Paraná mirror those of the broader Rondônia state: the lure of economic opportunities, particularly cattle ranching and soybean cultivation, has driven settlers to clear land, often through slash-and-burn methods. The construction of roads, such as the Highway BR-364, has also facilitated access to previously untouched areas, further accelerating deforestation. Over time, the landscape in Ji-Paraná is changing dramatically as forests are converted into agricultural fields and pastures. Loss of biodiversity, disruption of local ecosystems, and contribution to climate change are some of the negative effects of this environmental degradation.

Figures 6.5 and 6.6 show the land use change in the urban scale of Ji-Paraná. The Figures show how the forest areas have decreased from 1985 to 2021 and the urban areas have increased. The reason for this sharp decline of the forest areas is due to the above-mentioned reason.

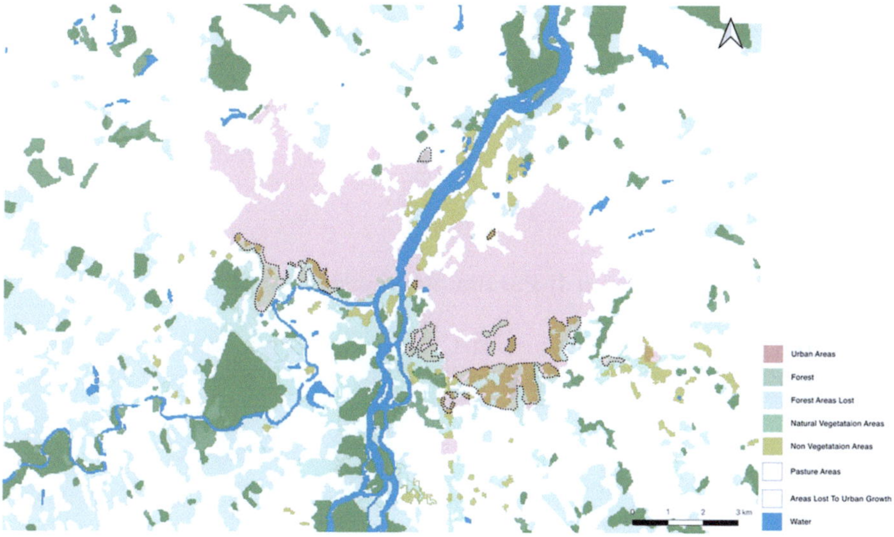

Fig. 6.5 Ji Paraná land use and land cover change from 1985 to 2005. (Source: Elaborated by Saeed, Tahir Ali, and Gül)

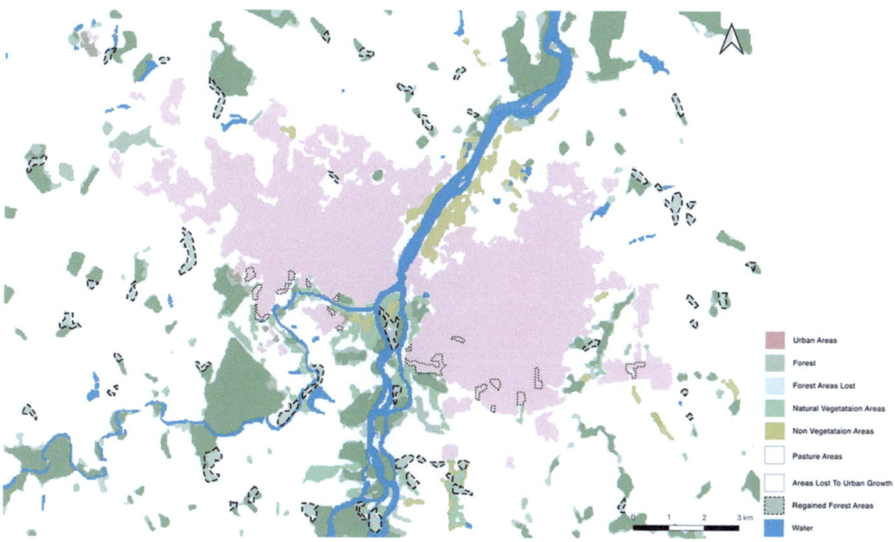

Fig. 6.6 Ji Paraná land use and land cover change from 2005 to 2021. (Source: Elaborated by Saeed, Tahir Ali, and Gül)

Fig. 6.6 shows the decline in the forest areas from 2005 to 2021. From the Figure, it can also be seen that there have been some efforts to restore the forest. However, forest loss has been greater than the areas where forest has been regained. Further explanation is given below.

6.3.4.1 Primary Forest Loss in Ji-Paraná, Rondônia

From 2002 to 2022, Ji-Paraná lost 15.9 kha of humid primary forest, making up 50% of its total tree cover loss in the same time period (Fig. 6.7). The total area of the humid primary forest in Ji-Paraná decreased by 3.6% in this time period (MapBiomas, 2023).

6.3.4.2 Tree Cover Loss in Ji-Paraná, Rondônia

From 2001 to 2022, Ji-Paraná lost 34.1 kha of tree cover, equivalent to a 6.8% decrease in tree cover since 2000, and 19.1 Mt. of CO_2 emissions, as shown in Fig. 6.8 (MapBiomas, 2023).

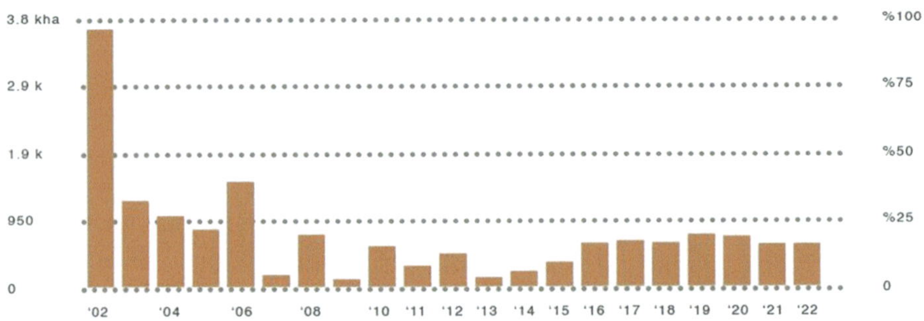

Fig. 6.7 Ji Paraná forest lost from 2002 to 2022. (Source: Elaborated by Saeed, Tahir Ali, and Gül, using data from Mapbiomas (2023))

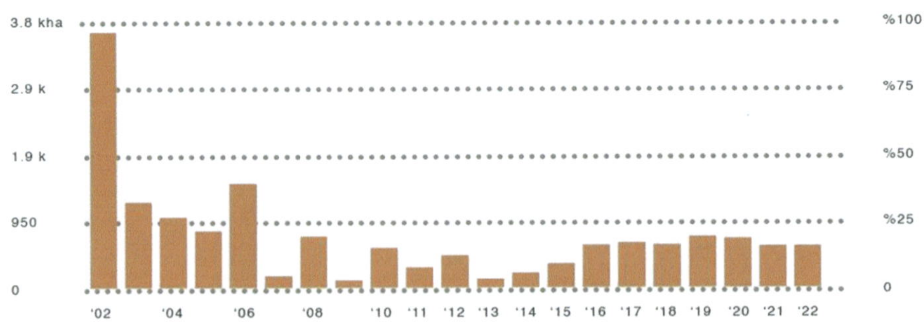

Fig. 6.8 Ji Paraná tree cover loss from 2001 to 2022. (Source: Elaborated by Saeed, Tahir Ali, and Gül, using data from Mapbiomas (2023))

6.3.4.3 Tree Cover Loss Due to Fires in Ji-Paraná, Rondônia

From 2001 to 2022, Ji-Paraná lost 2.56 kha of tree cover from fires and 31.6 kha from all other drivers of loss (Fig. 6.9). The year with the most tree cover loss due to fires during this period was 2017 with 451 ha lost to fires—32% of all tree cover loss for that year (MapBiomas, 2023). After our period of analysis, the situation became more critical: Greenberg (2022) highlighted August 2022 as the month with the highest number of fire hotspots recorded in the Amazon up to that year.

6.3.4.4 Components of Net Change in Tree Cover in Ji-Paraná, Rondônia

From 2000 to 2020, Ji-Paraná experienced a net change of −8.98 kha (−1.9%) in tree cover (Fig. 6.10).

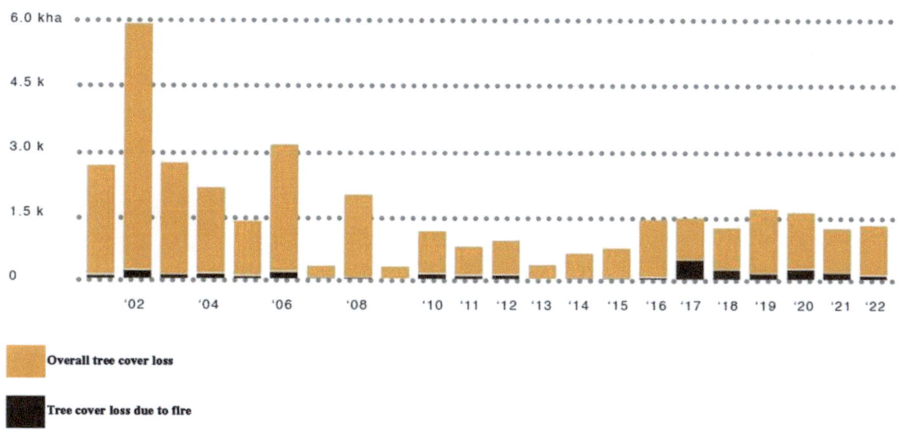

Fig. 6.9 Ji Paraná tree cover loss from 2001 to 2022. (Source: Elaborated by Saeed, Tahir Ali, and Gül, using data from Mapbiomas (2023))

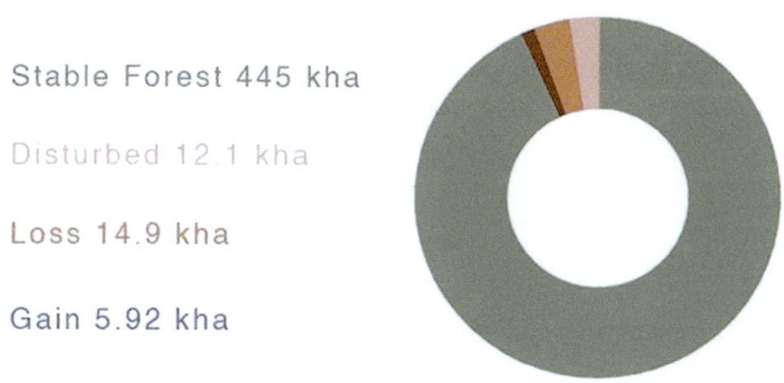

Fig. 6.10 Ji Paraná tree cover net change 2000–2020. (Source: Elaborated by Saeed, Tahir Ali, and Gül, using data from Mapbiomas (2023))

6.3.4.5 Tree Cover by Type in Ji-Paraná, Rondônia

As in 2000, 73% of Ji-Paraná land cover was >30% tree cover (Fig. 6.11).

6.3.5 Ji Paraná Land Use and Land Change Analysis

The land use data from MapBiomas (2021) and Esri (2017) shows how drastically the land use has changed in the state of Rondônia. This change can be seen from the Figs. 6.12, 6.13, and 6.14 and how the city of Ji-Paraná has expanded over the years from 1985 to 2021. It can be seen from the visualized Figures that urban growth is also playing its part in the deforestation around the city. The same trend can also be observed from 2005 to 2021.

Fig. 6.11 Ji Paraná tree cover by type 2000. (Source: Elaborated by Saeed, Tahir Ali, and Gül, using data from Mapbiomas (2023))

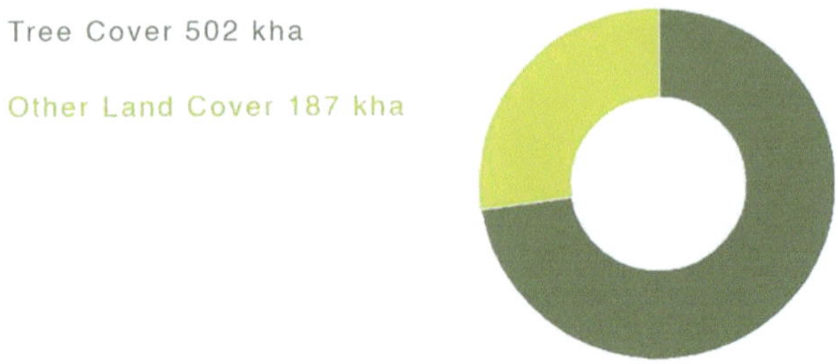

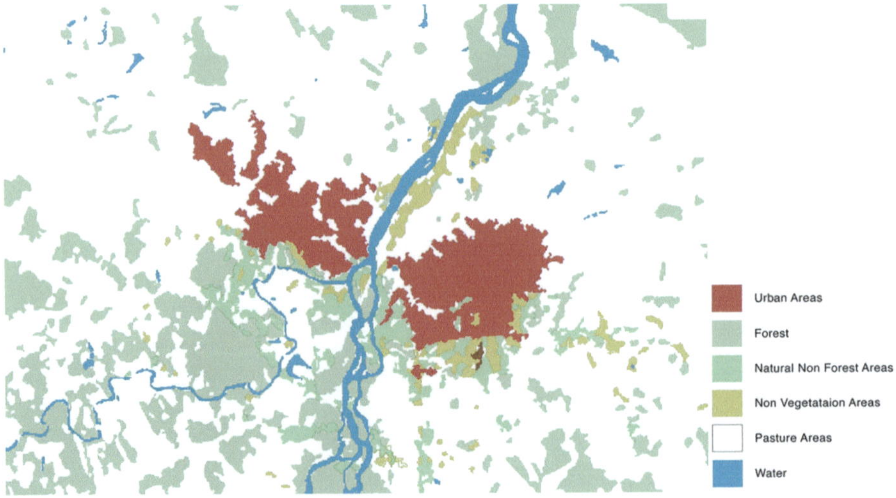

Fig. 6.12 Ji Paraná Land Use 1985. (Source: Elaborated by Saeed, Tahir Ali, and Gül, using data from Mapbiomas (2021))

Fig. 6.13 Ji Paraná Land Use 2005. (Source: Elaborated by Saeed, Tahir Ali, and Gül, using data from Mapbiomas (2021))

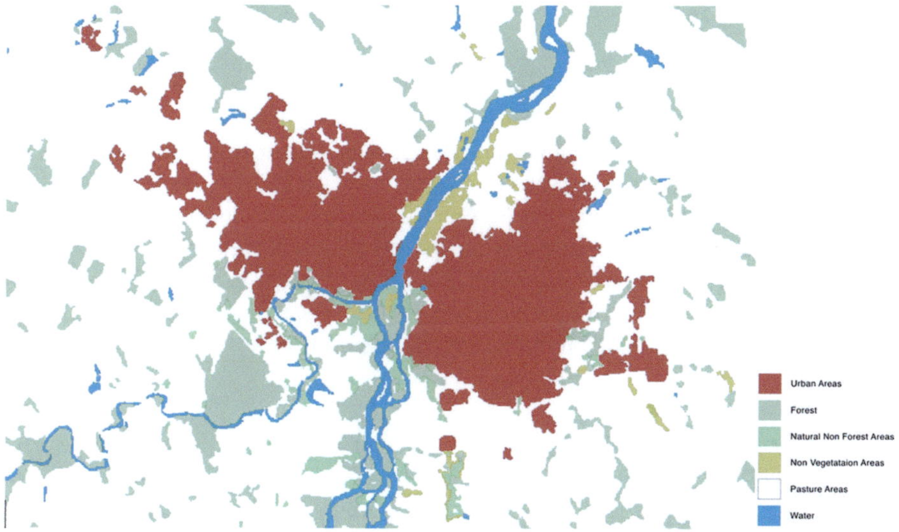

Fig. 6.14 Ji Paraná Land Use 2021. (Source: Elaborated by Saeed, Tahir Ali and Gül, using data from Mapbiomas (2021))

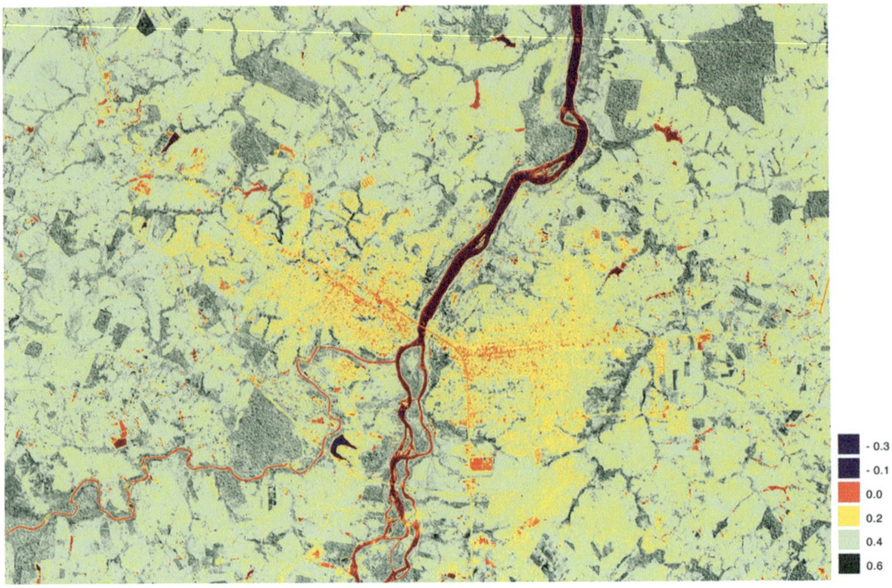

Fig. 6.15 Ji-Paraná NDVI Analysis. (Source: Elaborated by Saeed, Tahir Ali, and Gül)

6.3.6 NDVI Analysis

The land use and land cover change analysis provide information about the increasing urban areas and decreasing forest areas around the city of Ji-Paraná. The NDVI analysis was performed using the Sentinel 2 images and the raster calculator in QGIS. The resulting Figure shows where the vegetation is high and where it is low. It can be seen from the NDVI analysis that the vegetation in the city area is very low; this shows that the city lacks green public spaces and parks/gardens/open grounds for social gathering and biodiversity support (Fig. 6.15).

6.3.7 Precipitation Analysis

Brazil has an equatorial and tropical climate, with high annual rainfall. The country can be divided into five regions in terms of rainfall. Figures 6.15, 6.16, and 6.17 show that the northern and central regions of Brazil receive frequent rainfall, while the northeastern region receives less rainfall and has a semi-arid climate. The state of Rondônia is located in the northwest amazon basin. The region receives a good amount of rainfall throughout the year. Figs. 6.16, 6.17, and 6.18 present the information about the annual rainfall in Brazil in the year 2005, 2015, and 2021.

In the state of Rondônia the rainy season occurs mainly between January, February, March, April, October, November, and December. January is the wettest month in Rondônia, while July is the driest month. The average annual precipitation in Rondônia is 1928 mm (Figs. 6.19 and 6.20). This information is consistent with data provided by the Climate Knowledge Portal (2023).

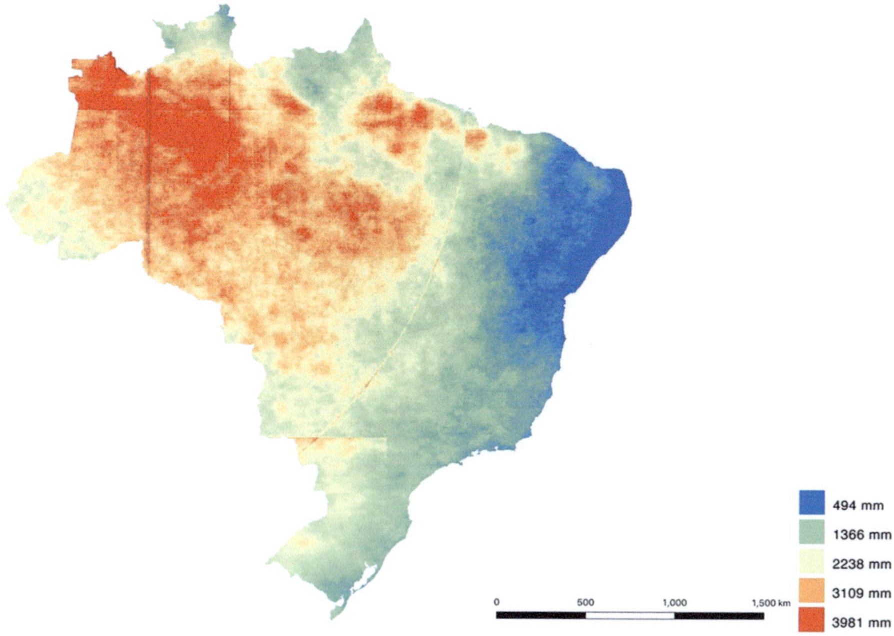

Fig. 6.16 Brazil Rainfall map 2005. (Source: Elaborated by Saeed, Tahir Ali, and Gül, using data from CHRS (2021))

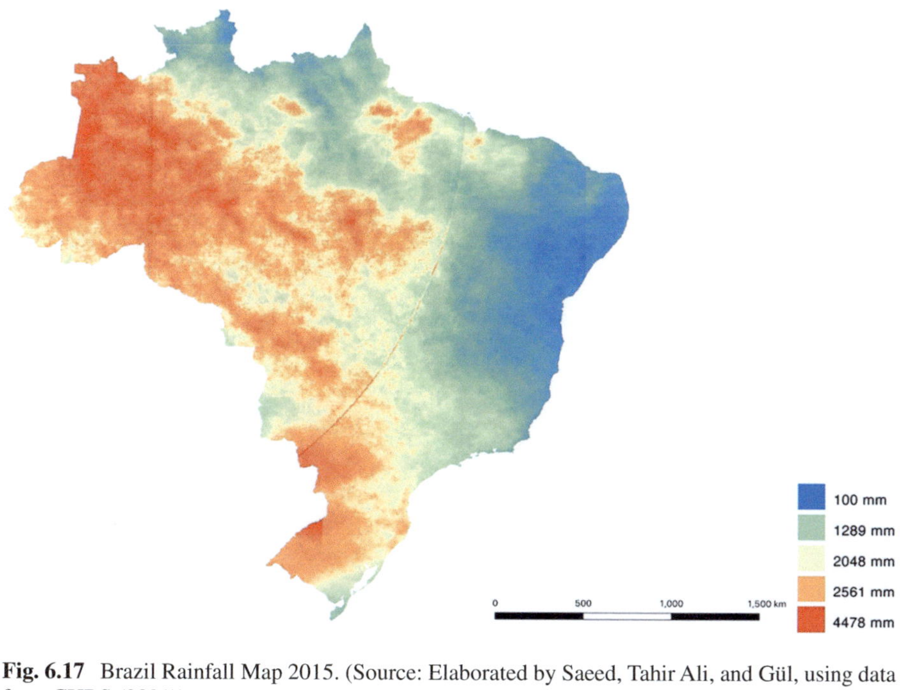

Fig. 6.17 Brazil Rainfall Map 2015. (Source: Elaborated by Saeed, Tahir Ali, and Gül, using data from CHRS (2021))

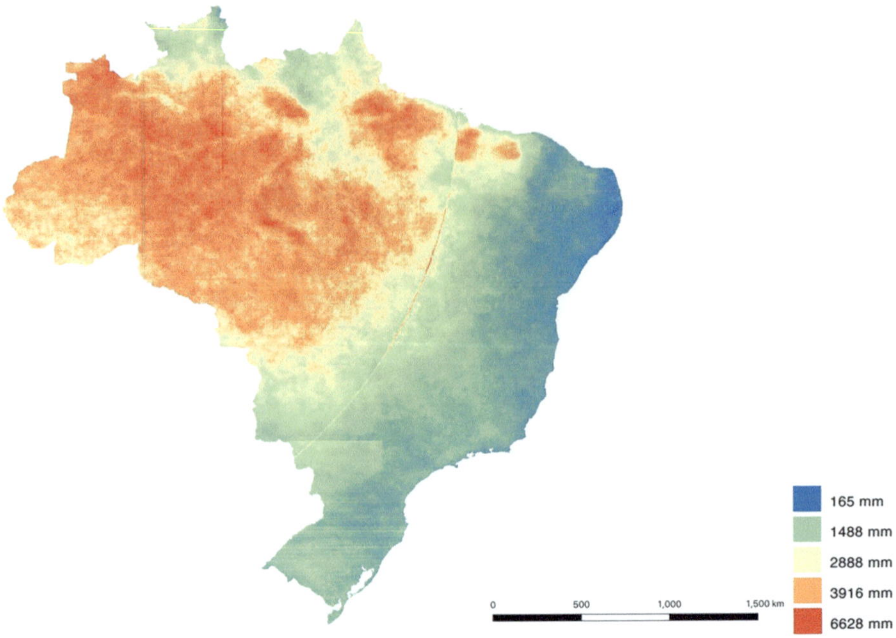

Fig. 6.18 Brazil Rainfall Map 2021. (Source: Elaborated by Saeed, Tahir Ali, and Gül, using data from CHRS (2021))

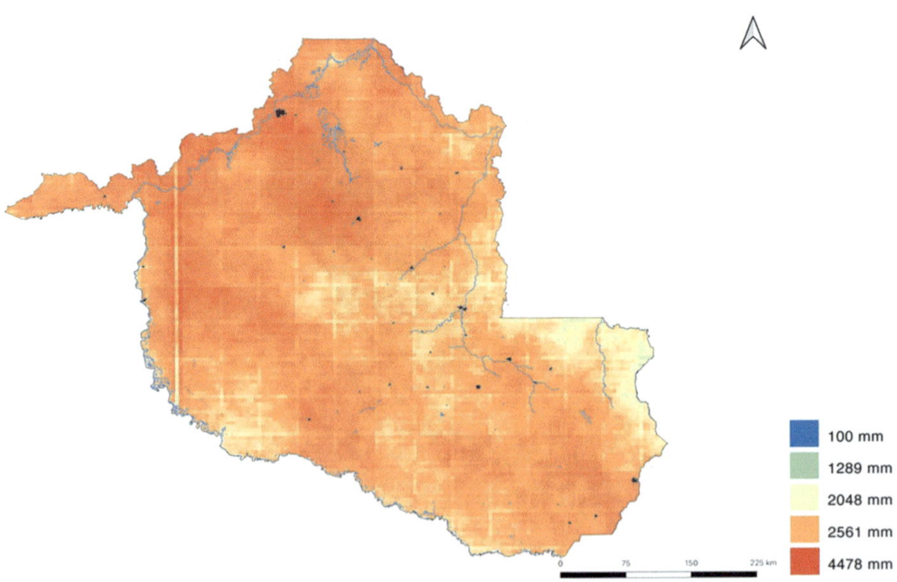

Fig. 6.19 Rondônia Rainfall Map 2015. (Source: Elaborated by Saeed, Tahir Ali, and Gül, using data from CHRS (2021))

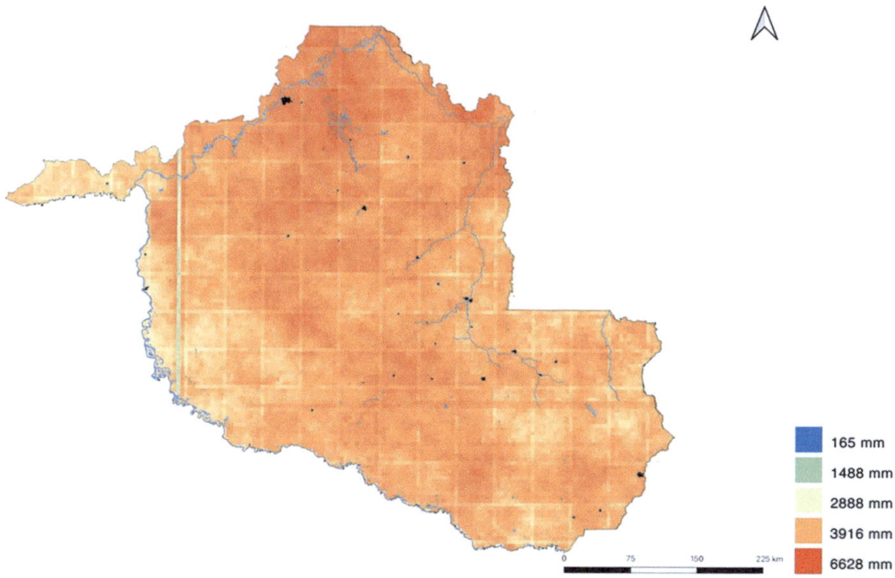

Fig. 6.20 Rondônia Rainfall Map 2021. (Source: Elaborated by Saeed, Tahir Ali, and Gül, using data from CHRS (2021))

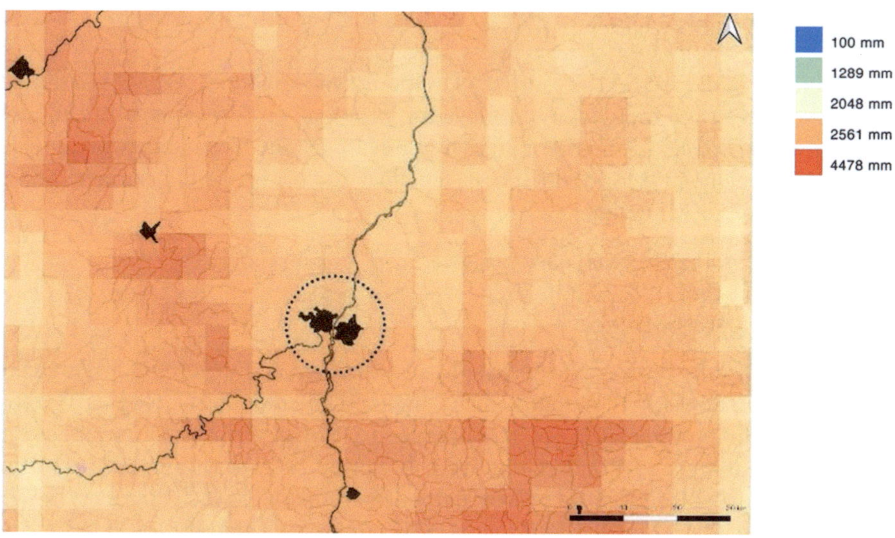

Fig. 6.21 Ji Paraná Rainfall Map 2015

Zooming in on the rainfall map analysis for Ji-Paraná (Figs. 6.21 and 6.22) it is possible to confirm that it has a tropical climate and receives an average of 1805 mm rainfall per year, with more rainfall in the summer and less in the winter.

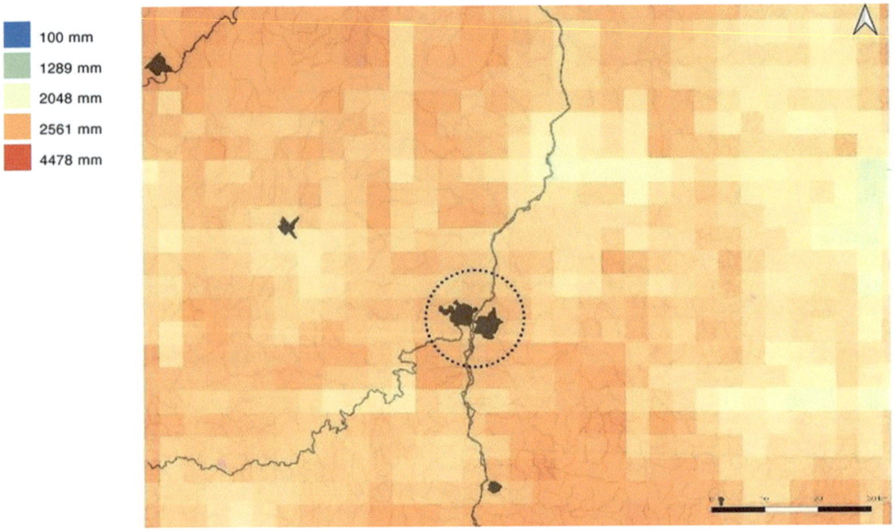

Fig. 6.22 Ji Paraná Rainfall Map 2021. (Source: Elaborated by Saeed, Tahir Ali, and Gül, using data from CHRS (2021))

6.3.8 Digital Elevation Model DEM Analysis

As described in the method session, the Digital Elevation Model (DEM) is a digital representation of the topography or elevation of a land surface. It is a crucial tool used in various fields for a wide range of applications due to its ability to provide accurate and detailed information about the elevation of the Earth's surface. Some applications of DEM include:

- Topographic mapping and cartography;
- Terrain analysis;
- -Hydrological modelling.

In order to study the terrain/topography of Rondônia and the city of Ji-Paraná, a DEM was created using the satellite data from the United States Geological Survey (USGS), Earth explorer website. The produced DEM gave information about the hydrological aspects of Rondônia as well. It can be seen from Fig. 6.23 that the northern region of Rondônia is mostly flat when compared to the central and south eastern region. Based on the DEM, it is also possible to identify the flood prone areas in the state and how much area can be potentially damaged when the water level in the rivers rise. Information on the Amazon River system was also obtained from Ziesler and Ardizzone (1979).

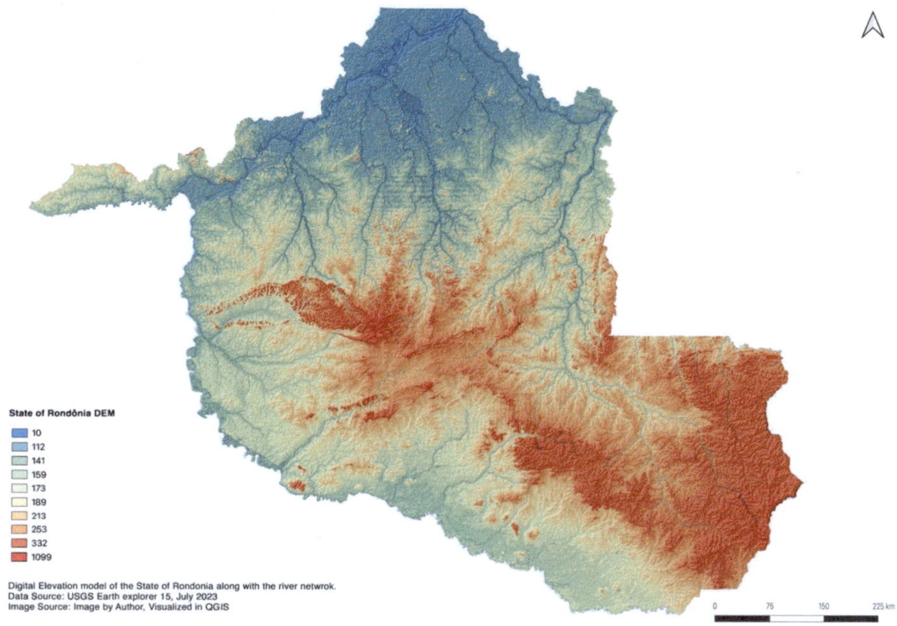

Fig. 6.23 Rondônia Digital Elevation Model. (Source: Elaborated by Saeed, Tahir Ali, and Gül)

6.3.9 Rivers of Rondônia

There are several rivers of Rondônia state. These are Madeira River, Mamoré River, Cautário River, São Miguel River, Branco River, Comemoração River, Roosevelt River, Jamari River, Candeias River, Jaci-paraná River (Fig. 6.24).

6.3.9.1 Ji-Paraná River—The Machado River

The Ji-Paraná River forms an important hydrological feature in the region, stretching for an impressive 900 km. Covering a basin of approximately 48,000 km^2, it flows through large areas of both the states of Rondônia and Mato Grosso.

The hydrological power of the river is further enhanced by its network of tributaries, which include the Comemoração River, the Anta River, and the São Lourenço River. These tributaries combine to contribute to the overall flow and volume of the Machado River.

Beyond its role as a waterway, the Ji-Paraná River serves as an important water source, providing a lifeline for irrigation and transportation activities. The river's waters are a veritable source of nourishment for local ecosystems, promoting biodiversity, and ecological balance. Moreover, the river's aquatic bounty extends to fisheries, harboring a variety of fish species that live in its depths.

In the field of agriculture, the Ji-Paraná River is important as a basic water source, meeting irrigation needs, especially in neighbouring lands. This hydrological

Fig. 6.24 Rondônia River Network—Ji-Paraná River. (Source: Elaborated by Saeed, Tahir Ali, and Gül)

lifeline sustains agricultural activities, strengthening the cultivation of crops and contributing to the agricultural vitality of the surrounding areas.

Ji-Paraná River emerges as a vital hydrological and ecological component. Its role ranges from providing water for human and agricultural needs to nurturing aquatic life, profoundly affecting both natural dynamics and human livelihoods.

6.3.10 Flood Analysis

As described previously, Rondônia receives a good amount of rainfall and it is also located in one of the wettest regions of Brazil. Therefore, it is prone to river and urban flooding. The river and flood hazard in the state of Rondônia is high, which means that there is a risk of life threatening and property damaging floods to occur at least once in 10 years (Thinkhazard 2023, Ariquemes 190, 2022, Rondoniagora, 2019, 2020, Rul a melhor informação, 2019, Canaa, 2016). Figures 6.25, 6.26, and 6.27 present some of the findings.

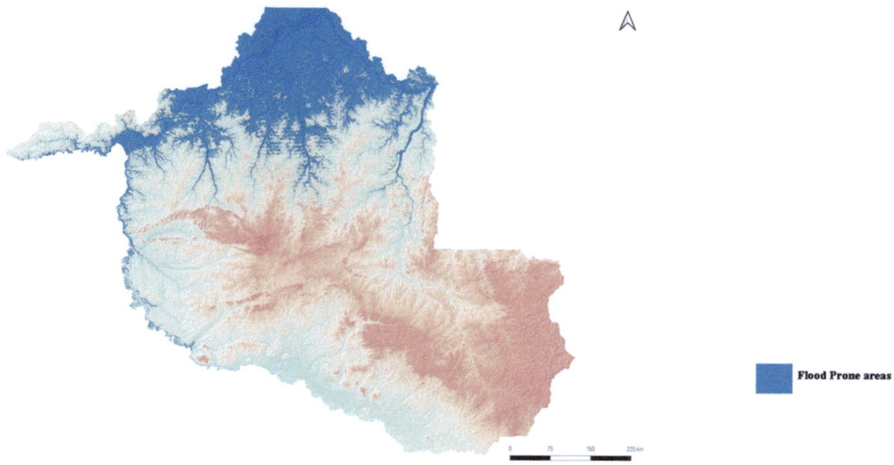

Fig. 6.25 Rondônia Flood prone areas with 10 m flood. (Source: Elaborated by Saeed, Tahir Ali, and Gül)

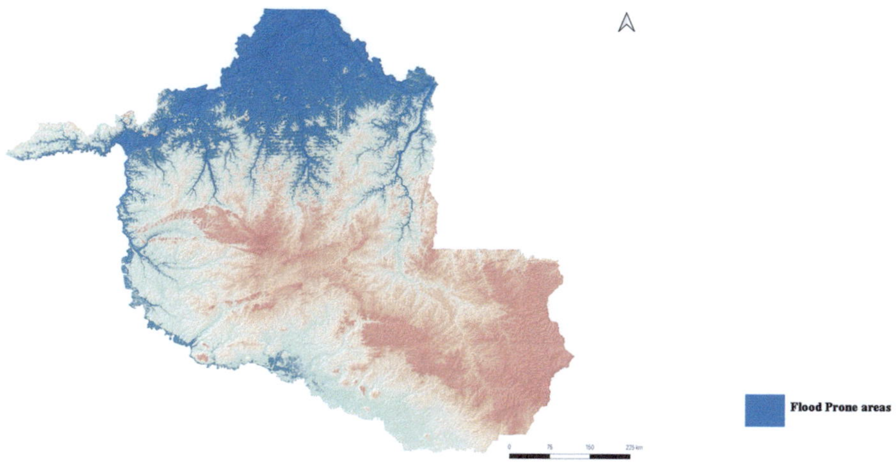

Fig. 6.26 Rondônia Flood prone areas with 20 m flood. Source: Elaborated by Saeed, Tahir Ali, and Gül

The same pattern goes for Ji-Paraná. The river and urban flooding are getting more frequent and severe. Rondônia was hit by one of the worst floods in 2014 due to heavy rainfall. The water levels of the Madeira River rose to 19.52 m and almost 20,000 people were affected (Davies, 2014).

Apart from this historic record, the most recent floods which have affected the Rondônia state and the city of Ji-Paraná occurred in 2022 where the flood level rose to 11.60 m in the Machado River. Almost 50 families were affected due to the floods (Ariquemes 190, Feb, 2022).

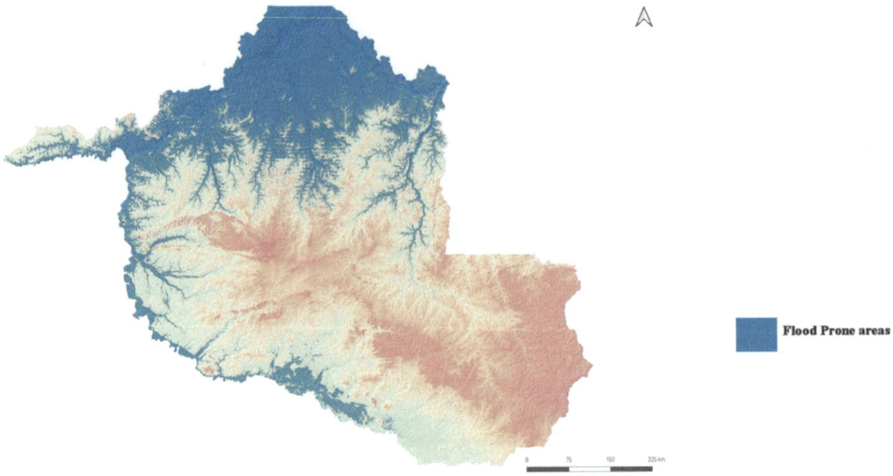

Fig. 6.27 Rondônia Flood prone areas with 30 m flood. (Source: Elaborated by Saeed, Tahir Ali, and Gül)

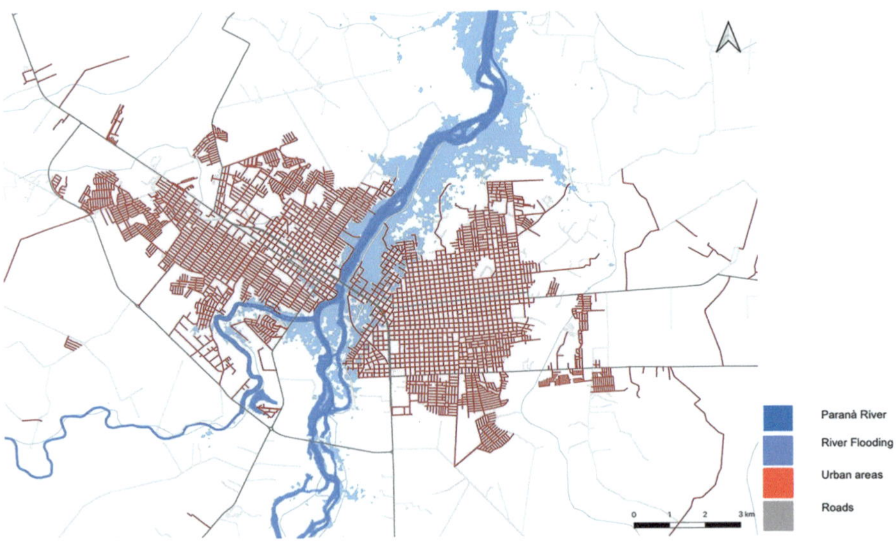

Fig. 6.28 Ji Paraná Flood prone areas with 20 m flood. (Source: Elaborated by Saeed, Tahir Ali, and Gül)

The DEM was used to identify the flood prone areas in the state of Rondônia and therefore create a flood risk analysis based on three different scenarios. Research data pointed that the highest flood rose to 19.52 m in 2014 in Rondônia (Davies, 2014). A raster calculation was made to identify the areas which are prone to flooding if the water level rises to 10 m, 20 m, and 30 m. Figs. 6.28, 6.29, and 6.30 show the flood prone areas on the state level and the city level, respectively.

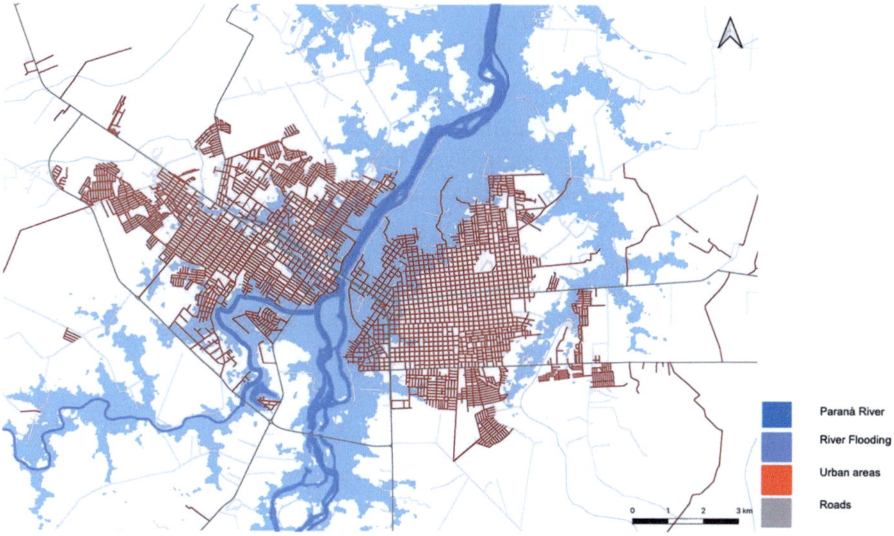

Fig. 6.29 Ji Paraná Flood prone areas with 30 m flood. (Source: Elaborated by Saeed, Tahir Ali, and Gül)

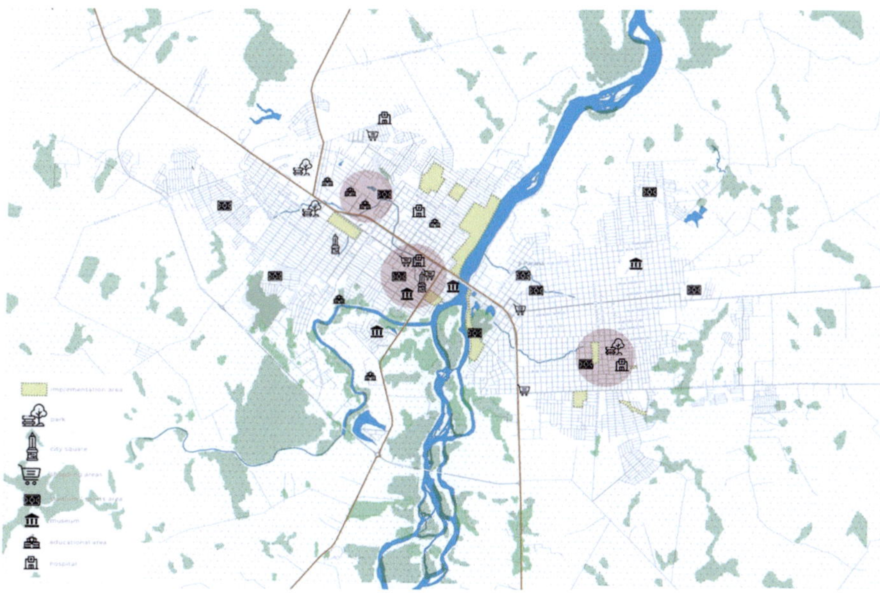

Fig. 6.30 Important points of the city. (Source: Elaborated by Saeed, Tahir Ali, and Gül)

6.3.11 Locations of the City

During the research phase, identifying the city's popular hubs was a challenge. Our methodology involved a mixed method using Google Maps and systematically exploring the city streets using Google Street View. As pointed at Fig. 6.30, it would be an oversimplification to say that there is a homogeneous distribution of open spaces and social hubs in the city. However, the scarcity of social spaces is also noteworthy. In a society whose culture is based on festivals, dance and music, the city lacks public spaces that allow different social groups to come together. The visual boundary in red defines the areas shown in Fig. 6.30 of social density in Ji-Paraná.

6.3.12 The Machado River

These include bustling commercial districts, shopping malls, cultural institutions, health facilities, recreational parks, educational institutions, and sports grounds. However, an important detail to note when studying the city of Ji-Paraná is that when we take a closer look at the areas marked and defined as parks in Fig. 6.31, it is possible to observe that most of them are soccer fields or areas with garbage disposal.

This unintentional limitation effectively restricts the accessibility of such social platforms to a specific social group. A closer look at the densely populated eastern

Fig. 6.31 Points around the river. (Source: Elaborated by Saeed, Tahir Ali, and Gül)

district reveals a concentration of commercial residential areas. However, looking closer at the shopping street it is possible to notice that the side land of the city's largest shopping centre is idle. This aspect shows us a crucial concept: any intervention in the city must go beyond mere cartographic representation and elevated perspectives.

In a city like Ji-Paraná, it is important to approach the human and street scale, because a top-down approach risks falling into the traps of a distant European urban template, devoid of local reality. Our approach during this study is a conscious strategy to align our interventions with the identity of Ji-Paraná. Recognizing that a successful adaptation depends on the nuances of this city, we aimed to guide each step of our implementation with care and a clear awareness of its characteristics.

While exploring the city, something that really caught our attention was the use of the riverfront. The Ji-Paraná River flows right through the middle of the city. In the heart of the city and by the river, it would be expected to see some social activities. However, this is not what the analysis shows for Ji-Paraná. People from the city do come to this area, but strangely, they don't really have much to do there, since the river area is not designed to be inviting.

6.3.13 Typology

See Fig. 6.32.

This grid system presented in Fig. 6.33 is particularly prevalent in densely populated residential areas. Street typologies 1, 2, and 4, which are mostly located on the periphery of the city, have a lower grid density. On the other hand, typologies 3, 5,

Fig. 6.32 Typology Map. (Source: Elaborated by Saeed, Tahir Ali, and Gül)

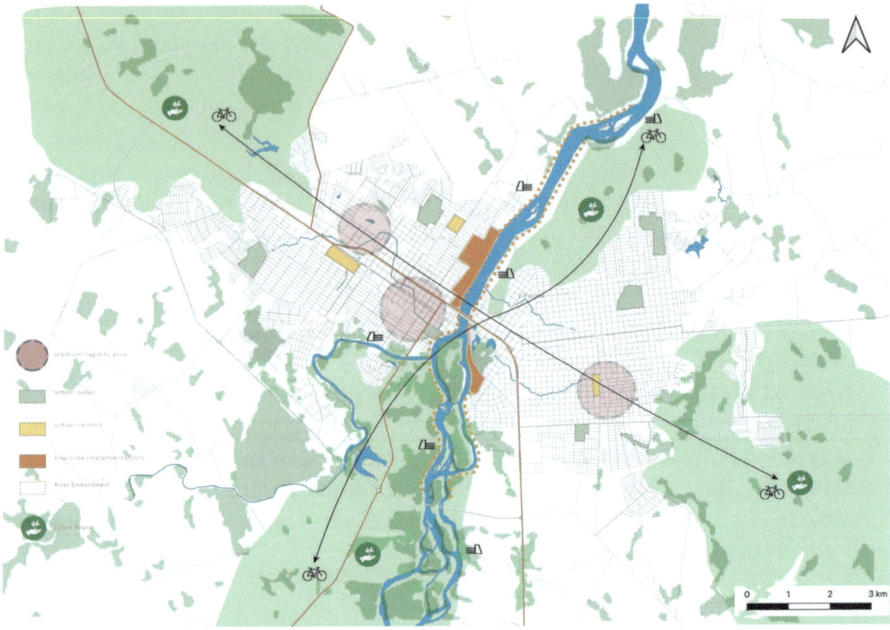

Fig. 6.33 Implementation areas map. (Source: Elaborated by Saeed, Tahir Ali, and Gül)

and 6 are mainly concentrated in densely populated areas. These typologies create a more uniform and regular urban arrangement in these areas.

Using the insights from these different street typologies, our study attempts to identify suitable areas within the application areas that can be designed as public spaces, parks, or squares. This strategic approach is consistent with the goal of harmonizing urban design with the existing urban context. Finally, to increase the functionality of Ji-Paraná's urban landscape.

6.4 Results

6.4.1 Implementation Areas

Figure 6.34 shows a proposal for the urban area where new parks and centres are to be created. These implementation areas have been selected according to the social density and needs of the city. At the same time, the red circles show these areas of intensive use and need. Around the river, which is one of the most important areas of the project, we have identified suitable areas where social spaces can be located, which we have examined through Google Maps. The pedestrian and bicycle paths that can be implemented in the city and along the river will significantly enhance mobility.

6.4.2 Strategies

6.4.2.1 Riverside Strategies

Parks and Public Spaces (Figs. 6.35 and 6.36)

- Creating recreational opportunities to engage social and community engagements;
- Creating pedestrian-friendly walking and cycling paths next to the riverside for people to interact with nature and green spaces inside the city;
- Creating urban furnitures, playgrounds, and picnic areas for social interactions;

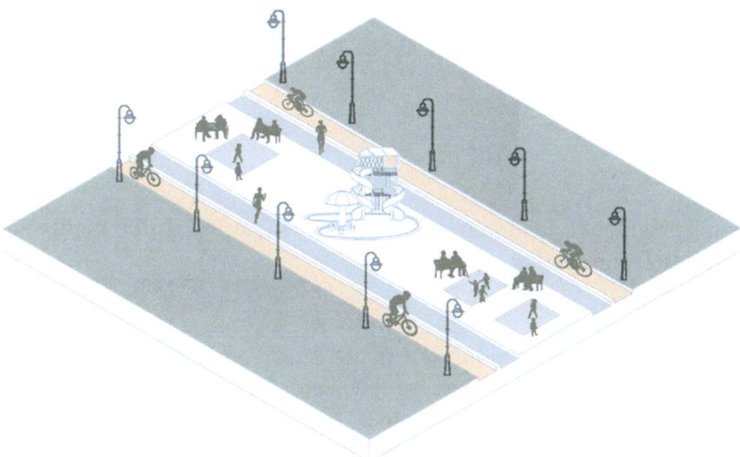

Fig. 6.34 Park

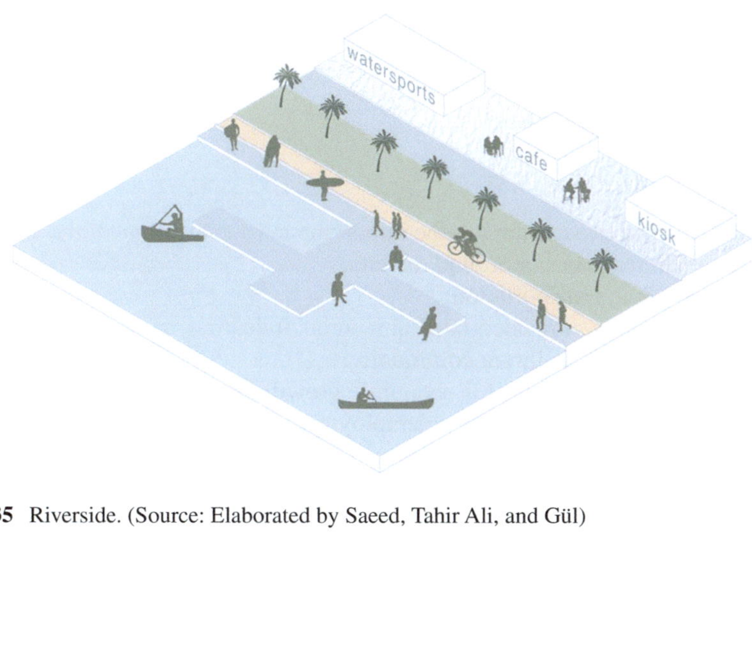

Fig. 6.35 Riverside. (Source: Elaborated by Saeed, Tahir Ali, and Gül)

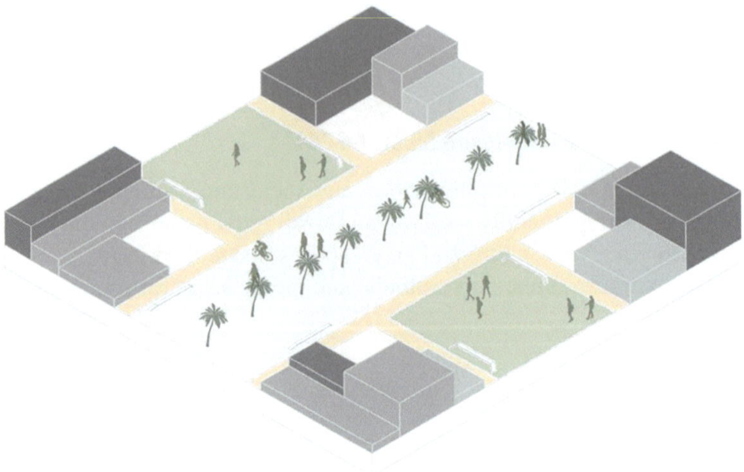

Fig. 6.36 Urban centre

- Designing attractive landscapes where people can use in the night safely time with lighting, and seating areas.

Social Interactions

- Mixed-use development along the riverside, creating social commercial areas like cafés, or kiosks for attraction of the people from the city and involving local communities on these areas for the local economy;
- Restaurants, cafes, and entertainment areas with waterfront views will help to build a vibrant scene, attracting visitors and locals.

Water-Based Activities

- Promoting water-based activities facilities such as boating, kayaking, or fishing.

6.4.2.2 Urban Level Strategies

New Urban Centres (Fig. 6.37)

- New urban centres according to the important and social points of the city where everyone can reach easily around the city;
- Accessible urban transportation;
- Encouraging a mix of residential, commercial, industrial, and institutional developments will create a vibrant community;
- New urban centres can act for economic growth, attracting investments, industries, and businesses with creating employment opportunities, new businesses to help to encourage the local economy;
- Creating of inclusive centres with social services, and community centres that facilitates social interaction.

Urban Parks (Fig. 6.38)

- Creating urban parks to the idle areas where there is no designed use in the area;
- Creating different recreational activities, not just football fields, other sports, such as basketball courts, volleyball courts, or tennis courts;
- Open green spaces for social activities, walking paths, jogging tracks, and areas to interact with nature;
- Redesigning idle areas with landscape design with integrating green infrastructure elements such as trees, shrubs, and gardens;

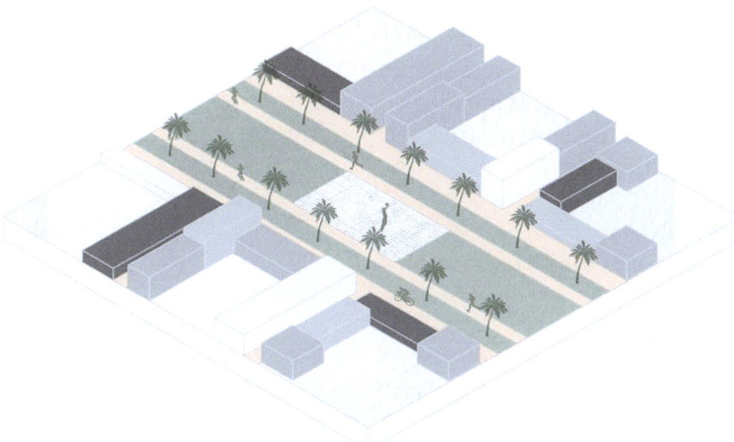

Fig. 6.37 Urban park. (Source: Elaborated by Saeed, Tahir Ali, and Gül)

Fig. 6.38 Riverside Vision. (Source: Elaborated by Saeed, Tahir Ali, and Gül)

- Creating urban furniture, playgrounds, exercise stations, outdoor fitness equipment;
- Connecting different green areas inside the city for the green connectivity inside the city.

6.4.2.3 Social Strategies

Cultural Areas

- Creating cultural hubs around the city where locals can develop art installations, galleries, performance spaces, and cultural events to create a vibrant and creative atmosphere, attracting residents;
- Cultural hubs, offering spaces for artistic expression, cultural activities, and the preservation of local heritage;
- Art galleries, museums, theatres, and performance venues that promote cultural diversity and creativity.

6.4.2.4 Strategies for Conservation

Protecting

Immediate elimination of deforestation: Enforce zero deforestation commitments in the beef and soy chains and implement legislation to prevent further habitat loss. Restoration efforts: Focus on key areas within the Amazon, targeting legally protected reserves and regaining connectivity with preserved areas.

Preserving

Preservation of connectivity: Preserve and restore connectivity between different areas of the Amazon to maintain biodiversity and ecosystem health. International cooperation: Collaborate with importing countries and organizations to promote sustainable practices and enforce regulations.

Rebuilding

Education and awareness: Raise public awareness about the importance of the Amazon, its biodiversity, and the need for conservation.

6.4.2.5 Flood Related Strategies

- River embankments near urban areas;
- Green infrastructure and sustainable drainage system;
- Early warning flood systems;
- River restoration and channel management;
- Flood Hazard management.

6.4.3 Future Visions

Figure 6.39 illustrates a prospective vision for the riverside area of the city using a collage method. The proposed alterations to the area are relatively small, involving the addition of a few structures to an otherwise unoccupied space. The most crucial element to be considered is the addition of a bicycle and pedestrian path. This can facilitate safer access for the general public. Furthermore, the proposal includes the introduction of water sports along the riverside, which has potential to become a significant attraction. This proposal intervention can have the effect of revitalizing the area socially, thereby encouraging its development as its use increases. In this area, which is currently characterized by a landscape of shrubs, dust and soil, a landscape arrangement has been proposed. Together with the integration of society and nature, the intention is to eliminate any sense of abandonment and to create a design that incorporates nature to the daily life.

Figure 6.39 depicts a vision for the most popular shopping streets in the city. The most striking remarks made during the analysis of the area were the absence of bicycle lanes and the inadequacy of street lighting. Thus began our initiative to

Fig. 6.39 Street Vision. (Source: Elaborated by Saeed, Tahir Ali, and Gül)

create a safe space for individuals who do not rely on automobiles. Even though it is one of the most popular streets in the city, we were intrigued by the absence of a defined road layout and the lack of designed social spaces. In order to encourage people to spend time on the street, small pop-up markets and urban furniture were designed. The lack of greenery in the city was also very striking, and that's why the addition of palm trees on the sidewalks of the streets was considered, for this particular street.

6.5 Conclusion

In conclusion, this chapter provides a comprehensive analysis of environmental challenges in Rondônia, with a focus on Ji Paraná. The data-driven approach reveals alarming deforestation rates, driven by factors like agriculture and logging. This trend also affects Ji Paraná, where urban growth contributes to deforestation. The study highlights the need for sustainable urban planning and preservation of green spaces. Insights from NDVI, precipitation, and temperature patterns underscore environmental vulnerabilities, emphasizing the importance of proactive strategies. By bridging regional issues with local impacts, the chapter advocates for collaborative efforts among communities, policymakers, and organizations to mitigate deforestation, protect biodiversity, and enhance resilience. In social strategies urban landscape, providing residents with spaces for leisure, culture, and community interaction. The implementation of these interventions not only enhances the physical environment but also contributes to economic growth, social cohesion, and the overall well-being of the city's inhabitants. This analysis serves as a valuable resource for informed decision-making, offering pathways for sustainable development and environmental conservation.

References

Ariquemes 190. (2022). *Rio Machado sobe mais e Ji-Paraná emite alerta vermelho*. Retrieved June 2023, from https://ariquemes190.com.br/noticia.php?id=64585

Boulton, C. A., Lenton, T. M., & Boers, N. (2022). Pronounced loss of Amazon rainforest resilience since the early 2000s. *Nature Climate Change, 12*, 271–278. https://doi.org/10.1038/s41558-022-01287-8

Canaa, R. (2016). *Ji-Paraná - Após 5 horas de muita chuva, igarapé Dois de Abril transborda e deixa várias famílias desabrigadas*. Retrieved July 2023, from http://www.canaafm104.com.br/noticias/noticia/id:80;ji-parana-apos-5-horas-de-muita-chuva-igarape-dois-de-abril-transborda-e-deixa-varias-familias-desabrigadas.html

Climate Knowledge Portal. (2023). *Brazil climatology*. Retrieved July 2023, from https://climateknowledgeportal.worldbank.org/country/brazil/climate-data-historical#:~:text=Dominated%20by%20equatorial%20and%20tropical,mm%20per%20year%20of%20rain

Copernicus. (2023). *European Union's Earth observation programme*. Retrieved August 30, 2023 from https://www.copernicus.eu/en
da Cunha, E. T., & Moser, L. M. (2010). Os projetos de colonização em Rondônia. Dossiê "A construção do Sujeito Contemporâneo: Perspectiva para a Amazônia—parte II". *Revista Labirinto, X*, 14.
Davies, R. (2014). *20,000 Hit by Floods in Rondônia, Brazil*. Retrieved July 2023, from https://floodlist.com/america/floods-rondonia-brazil
Earth Observatory.nasa.gov. (2023). *Deforestation in Rondonia, Brazil*. Retrieved June 2023, from https://earthobservatory.nasa.gov/images/7548/deforestation-in-rondonia-brazil
Fearnside, P. and Cruz, P. V. (2018). *Chainsaw massacre: Protected areas in danger in Brazil's state of Rondônia (commentary). Mongabay, 31 October*. Retrieved August 30, 2023, from https://news.mongabay.com/2018/10/chainsaw-massacre-protected-areas-in-danger-in-brazils-state-of-rondonia-commentary/
Forest champions (2020) *Rondônia Brazil, Jurisdictional Indicators brief, BR-RO-10*. Retrieved June 2023, from https://forestchampions.org/jxd_reports/en_Rond%C3%B4nia_Brazil.pdf
GISGeography. (2023). *What is NDVI (Normalized Difference Vegetation Index)?* Retrieved August 30, 2023 from https://gisgeography.com/ndvi-normalized-difference-vegetation-index/#:~:text=As%20show%20below%2C%20 Normalized%20 Difference,vegetation%20 as%20the%20color%20green
Global Forest Watch. (2023). *Brazil, Rondônia*. Retrieved June 1, 2023, from https://www.globalforestwatch.org/dashboards/country/BRA/22/?category=undefined
Greenberg, C. (2022). Amazon rainforest fires 2022: Facts, causes, and climate impact. Retrieved July 2023, from https://www.greenpeace.org/international/story/55533/amazon-rainforest-fires-2022-brazilcauses-climate/
IBGE. (2023). *Brazilian Institute of Geography and Statistics*. Retrieved August 30, 2023, from https://cidades.ibge.gov.br
Jacques Imbernon. (n.d.). (no date). P*opulation growth and deforestation— the case of Rondonia in the Brazilian Amazon*. Retrieved June 2023, from https://apps.worldagroforestry.org/Units/Library/Books/PDFs/80_ASB_technical_report_no_3.pdf
MapBiomas. (2021). *Collection 7*. Retrieved June 2023, from https://plataforma.alerta.mapbiomas.org/
Mapbiomas. (2023). Brazil annual land use and land cover mapping project. *Collection 8.0*. Retrieved June 2024, from http://www.mapbiomas.org
Neves, E. G. (2022). Sob os tempos do equinócio. In *Oito mil anos de história na Amazônia central*. Editora USP.
Rondoniagora. (2019). *Quase 60 famílias são retiradas de casa por causa da cheia do Rio Machado, em Ji-Paraná*. Retrieved July 2023, from https://www.rondoniagora.com/cidades/quase-60-familias-sao-retiradas-de-casa-por-causa-da-cheia-do-rio-machado-em-ji-parana
Rondoniagora (2020) *JI-PARANÁ: Rio Machado passa dos 10 metros e defesa civil alerta para nova enchente*. Retrieved July 2023, from https://rondoniaovivo.com/noticia/interior/2023/03/22/ji-parana-rio-machado-passa-dos-10-metros-e-defesa-civil-alerta-para-nova-enchente.html
Rul A Melhor Informação (2019). *Jornalista registra imagens da enchente em Ji-Paraná*. Retrieved July 2023, from http://www.rul.com.br/noticias-detalhes.php?cod=17589
SPA (2021). *Science Panel for the Amazon*. Amazon Assessment Report 2021. Retrieved September 2023, from https://www.theamazonwewant.org/wp-content/uploads/2022/06/220717-SPA-Executive-Summary-2021-EN.pdf
WWF. (2021). *Destroying native vegetation to produce beef and soy is the main driver affecting biodiversity in the Carrado and amazon*. Retrieved September 2023, from https://www.wwf.de/fileadmin/fm-wwf/Publikationen-PDF/Amazonas/WWF-NotaTecnica-Desmate-e-perda-de-especies-2021-english.pdf
Ziesler, R., Ardizzone, G.D. (1979). *Amazon River System*. 2013.

Chapter 7
Ariquemes, the 15-Minute City: A Conceptual Proposal to Strengthen Social Inclusion and Sustainable Greening Through a 15-Min Urban Planning System

Shristi Thakuria and Deniz Özdemir

Abstract Ariquemes, located in the heart of the Brazilian Amazon in the state of Rondônia, faces a complex set of urban challenges that require innovative solutions. Historically plagued by deforestation, rapid land use change and infrastructure deficiencies, the city is grappling with the consequences of unbridled urbanization. This book chapter explores a proposed transformative urban planning system, the "*15-Minute City*," tailored for Ariquemes. This concept seeks to reconfigure urban morphology and promote social inclusion, while addressing the environmental consequences of deforestation. By weaving accessible green spaces and essential services into the urban fabric, the concept aims to create a resilient, inclusive, and sustainable urban landscape. It is an integrative methodology that combines urban planning, community engagement, and sustainability principles. The results demonstrate the potential for the 15-Minute City planning in Ariquemes as a catalyst for social inclusion and green living. This project not only demonstrate the city's ability to address these complex issues but also highlights the imperative of respecting Indigenous/minority rights and cultural heritage in the context of planning. The "Ariquemes 15-Minute City" stands as a beacon for the urban future through the adoption of the 15-Minute City concept for the urban planning system, proving that resilient and green urban living is within reach, even in the heart of the Amazon.

Keywords Ariquemes · Rondônia · Sustainable development · Urban planning · Integrative planning

---

S. Thakuria (✉) · D. Özdemir
RWTH Aachen University, Aachen, Germany
e-mail: shristi.thakuria@rwth-aachen.de

V. G. Donoso, C. Reicher (eds.), *Rethinking Amazonian Cities in the State of Rondônia*, The Latin American Studies Book Series,
https://doi.org/10.1007/978-3-032-01817-5_7

7.1 Introduction

The Amazon rainforest, often referred to as the "lungs of the Earth," is the largest tropical rainforest in the world. It stretches across nine countries in South America, with a majority of its territory located in Brazil. Within Brazil, the state of Rondônia, located in the western part of the Brazilian Amazon, holds unique importance due to its particularities. Rondônia has experienced significant deforestation and land use change, driven by agricultural expansion, logging, and infrastructure development. Its complex socio-environmental landscape highlights the tensions between conservation efforts and economic development in the Amazon.

Rondônia's history is intertwined with that of its Indigenous populations, and the state has been a focal point for discussions on indigenous rights and cultural preservation. The complex dynamics of this region serve as a microcosm of the broader challenges facing the Amazon and illustrate the delicate balance required to protect its ecological richness while meeting the socio-economic needs of its inhabitants (Nepstad et al., 2006; Fearnside, 2005).

According to the World Wildlife Fund (WWF, 2023), deforestation in the Amazon was three times higher in the first half of 2022 than in the first half of 2017. It has increased every year for the past 5 years, with no sign of slowing down.

The MapBiomas (2021) project has analyzed land use change in Brazil over the last 38 years using satellite imagery. Their data show the conversion of natural areas, such as forests and grasslands, to agriculture and pasture over this period. Most of Brazil's recent deforestation due to land use change for pasture has occurred in the Amazon region, and Rondônia state has played a key role in this story. Rondônia is located in the south-western part of the Brazilian Amazon and was originally entirely rainforest. In 2021, compared to 1985, the natural forest vegetation area in the state was reduced by 66% (Mapbiomas, 2021). The threats are related to land use changes that directly cause deforestation at the local or regional level. These include certain types of agriculture, mining, other extractive industries (e.g., oil, timber) and infrastructure that directly replace forests (Hänggli et al., 2023).

Rondônia is currently undergoing a period of major agricultural expansion, with the main crops being rice, banana, coffee, cassava, and soy. Besides the agricultural development, beef production is the main driver of deforestation in the Brazilian Amazon. However, its development has presented specific characteristics and challenges in terms of the occupation and use of the territory, as it has been marked by periods of different land use for agriculture, cattle ranching, and the exploration of mineral and environmental resources (Carvalho & Delgado, 2018).

Pedlowski et al. (1997) and Piontekowski et al. (2014) explain that the two main cycles of urban growth and development in the region were due to the nineteenth Century events and the World War II, the former being especially related to rubber extraction and the later to the exploitation of cassiterite (tin oxide, Sn O_2). These economic cycles were factors that increased the migration flow to the region.

However, these are not the only two main cycles in the growth of cities in the state of Rondônia, and the infrastructure development also plays a key role.

In 1985 the 364 Highway was built to connect Rondônia to the state of Cuiabá. This infrastructure encouraged the start of deforestation in Rondônia: Following the pavement of the highway in 1985, 8% of the region was deforested for farming. By 2017, 33% of the state's forested land had been converted to pasture and agriculture (MapBiomas, 2021).

Due to the highway construction, urbanization along the highway axis increased, also linked to the land grants made by the Institute of Colonization and Agrarian Reform (INCRA). The demarcation of settlements has led to a large flow of migration to the region, which has resulted in the expansion of the agricultural frontier and, on the other hand, has contributed greatly to the destruction of the forest (Tourneau & Bursztyn, 2010).

At the end of the 1970s, as a result of the Brazilian government's strategy of occupation and local development, implemented by INCRA, native forest areas were replaced by agriculture (temporary and permanent crops) and pastures for cattle farms (Tourneau & Bursztyn, 2010). Cardille and Foley (2003) confirm that in 1990, Rondônia had the highest deforestation indexes due to a large expansion of cattle ranching, with peaks in the municipalities of Ji-Paraná and Ariquemes. This expansion took place around Porto Velho and in areas in the south of the state.

Established in the late 1970s, Ariquemes' origins are linked to Brazil's National Integration Plan (PIN), a visionary undertaking plan orchestrated by INCRA. This aimed to unlock the Amazon's vast potential for colonization and agricultural expansion, propelling the city into existence (Nepstad et al., 2006).

Yet, as Ariquemes emerged, it became a microcosm of challenges facing the Amazon. The city's rapid development was paralleled by the reality of deforestation and dramatic land use changes, emblematic of the broader Amazonian narrative. While agricultural ambitions flourished, this growth came at an environmental cost, casting a shadow over the region's natural resources (Fearnside & Salati, 1985).

Ariquemes is the third largest population center in the state of Rondônia, with around 111 thousand inhabitants in 2022, according to the estimates of the Brazilian Institute of Geography and Statistics (IBGE, 2023a, 2023b). It is a nucleus of agricultural colonization commissioned by the INCRA, implanted on the banks of the Jamari River and close to a small pre-existing village. It was designed by the architect and urban planner Antonio Carlos Cabral Carpintero in the 1970s (de Paiva & Trevisan, 2023).

The topography and the pre-existing layout of the Highways BR-364 and RO-257 strongly influenced the position chosen for the urban core, which could not be a direct expansion of the old core, with the new city being built to the north. Carpintero also implemented two parallel roads about 2.5 km from the BR-364 Highway that functions as a structural axis of the urban morphology. These avenues delimit the Institutional Sector of the city, which has a different form of occupation, with a predominance of open spaces. This forms a different landscape in the midst of the urban layout of the city (de Paiva & Trevisan, 2023).

Perpendicular to the Institutional Sector, the connecting roads that delimit the neighborhoods of Ariquemes depart, originally called Sectors, with most shops and services placed close to the Institutional Sector. The blocks are rectangular in shape and are laid out sequentially as a regular plot (de Paiva & Trevisan, 2023). Currently, the city has been faced with many challenges pertaining to deforestation, severe land use change, lack of infrastructure, and so on, which we will analyze further in-depth.

To explore this complex reality, this chapter begins with an overview of the historical context of the city of Ariquemes. It then examines the impact of agricultural colonization and migration to the region from the 1970s onward, which affected the region's demography, culture, and socio-economic landscape.

This book chapter also explores a comprehensive approach to green urban planning and design. Rethinking the city offers us a unique opportunity to reconcile environmental conservation, Indigenous heritage protection, and sustainable development. By adopting sustainable practices, engaging Indigenous communities, promoting cultural preservation, and forging collaborations, we aim to create a better life in the city. It is within this framework that we answer our research questions: *(1) How can we re-imagine Ariquemes as a model city that thrives in harmony with its natural environment and celebrates its diverse cultural heritage? (2) How can we reduce the impact that agricultural expansion has on the ecology and environment of Ariquemes without hindering the economic benefits produced by the city? (3) How can we integrate the lost heritage and culture of the Indigenous communities and minorities of Rondônia?*

To further elaborate this study, the following **hypothesis** is proposed: *The implementation of sustainable urban planning strategies, including the integration of green infrastructure, efficient transport systems, adequate blue infrastructure, and cultural preservation initiatives, will lead to improved environmental quality, enhanced social well-being, and increased economic opportunities in Ariquemes.* Then, a **vision** is developed to strengthen inclusivity and promote sustainable greens, using the *15-min grid concept, a city planning strategy* and a strategic roadmap to guide the next steps of urban planning and design in the city. Finally, to conclude our research, an intervention is proposed in the form of a policy and a master plan.

7.2 Methodology

To answer our research questions, we systematically reviewed the available literature on Amazonian deforestation, agricultural colonization, the history of Ariquemes, and its socio-economic growth and development, from the years 1970 to 2023. We analyzed this evidence base using a framework derived from our theoretical model to identify specific deforestation drivers and how the city of Ariquemes has been affected by it.

The following steps were conducted:

7.2.1 Analysis

A comprehensive analysis of the city was conducted at both the macro- and micro-scales. This involved gathering data on the existing urban fabric, land use patterns, infrastructure, mobility, green spaces, and social nodes. This was followed by an assessment of the current strengths, weaknesses, opportunities, and challenges (SWOT analysis) within the area. This analysis provided a holistic understanding of the context and formed the basis for the following steps.

The analysis phase was followed by the integration of geospatial tools and diverse data sources, such as QGIS software for generating NDVI, NDBI, flood hazard, and elevation maps. This is complemented by the MapBiomas data source (2021) and satellite imagery for deforestation and land cover insights, and with data from the Brazilian Institute of Geography and Statistics (IBGE, 2023a, 2023b) for social analyses.

To calculate the NDVI and NDBI of specific time periods 1997, 2012, and 2023 we used satellite imagery from Landsat-TM (for 1997 and 2012), and Landsat-8/9 OLI (for 2023). From the NDVI and NDBI maps, we can summarize that the quantity of primary vegetation has been decreasing rapidly over time, mostly due to mass agriculture and urbanization of land. Also, the quality of bare soil and the agricultural lands itself have been deteriorating over time.

7.2.2 Problem Definition

Based on this analysis, the key urban design challenges are defined. These can be divided into two distinct categories: physical problems and socio-economic problems. This step forms the basis for the formulation of strategic interventions.

7.2.3 Strategic Planning

A comprehensive strategic plan has been developed to address the identified issues and opportunities, to create a vision for a vibrant and sustainable urban environment for the city. This is to be achieved by combining the 15-minute city and mixed-use functional concepts with the grid system as well as the integration of the green spaces into the city.

7.2.4 Strategy Framework

A strategy framework outlining the specific measures and actions required to implement the strategic plan was created. This consisted of three categories of improvement: environmental, infrastructural, and social. These improvements were later applied to five distinct layers: infrastructure, commercial, residential, green, and social.

7.2.5 Toolbox Component Selection and Application

A series of toolbox components, in line with the strategic plan and the problem definition, were identified and applied to the urban design area. The process was concluded with a master plan. Architectural and urban design-oriented software such as AutoCAD, Adobe Photoshop, QGIS, and so on, were used to conceptualize and create the graphics for the Master Plan, Strategic Map, Vision Map, and many more.

7.2.6 Timeline

A timeline was created to outline the steps and milestones of the planning and design interventions, including the future developments that could take place in the city. The timeline acts as a guide for execution of public policies as well.

7.3 Analysis and Discussion

7.3.1 Macro-Scale Analysis

7.3.1.1 Deforestation

The Amazonian tropical forests have been disappearing at a fast rate in the last 50 years due to deforestation to open areas primarily for agriculture, posing high risks of irreversible changes to biodiversity and ecosystems (MapBiomas, 2021). The rural development in the Amazonia pushed the agricultural frontier swiftly, resulting in widespread land cover change. However, agriculture in the Amazon has been of low productivity and unsustainable (Nobre et al., 2016).

The loss of biodiversity and continued deforestation may lead to high risks of irreversible change in its tropical forests. Modeling studies have shown that the Amazon may have two "tipping points," namely a temperature increase of 4 °C or deforestation of more than 40% of the forest area. If these thresholds are crossed, a large-scale "savannization" (change of forest type from rainforests to savannahs) may occur, mainly in the southern and eastern Amazon. This is alarming information, especially given that the region has warmed about 1 °C over the last 60 years, and total deforestation has reached 20% of the forested area (Nobre et al., 2016).

From 2002 to 2022, Rondônia lost 3,740,000 ha of humid primary forest, which is around 80% of its total tree cover, lost in the same period. The total area of humid primary forest in Rondônia decreased by 24% in this time (MapBiomas, 2021). However, the recent significant reductions in deforestation—an 80% reduction in the Brazilian Amazon in the last decade opens opportunities for a novel sustainable development paradigm for the future of the Amazon (Nobre et al., 2016).

According to the data collected from MapBiomas, the city of Ariquemes had an area of 347,296 ha of primary vegetation, that is, undisturbed plant community that develops in an area over a prolonged period in the absence of significant human or natural disturbances like fire or clear-cutting. This vegetation is in a stable state and is often well-adapted to the local climate and soil conditions.

The city also had an area of 150 ha of secondary vegetation, that is, plant community that develops in an area after it has been disturbed or altered in some way. This disturbance can be caused by human activities like agriculture, logging, or urban development, as well as natural events like wildfires or volcanic eruptions.

However, in 2000 this figure was estimated to be 223,150 ha of primary vegetation and 12,643 ha of secondary vegetation. Subsequently, we can observe even more decline in 2021 with only 134,635 ha of primary vegetation and 18,731 ha of secondary vegetation. When further investigated, the secondary vegetation included pasture, agriculture, forest plantation, beach, dune and sand spots, urban areas, mining, and other non-vegetated areas, with agriculture primarily constituting temporary crops like soybean, sugar cane, rice, cotton, etc. and perennial crops like coffee, citrus, and others (MapBiomas, 2021).

7.3.1.2 Land Cover Change

Land development in the Brazilian Amazon has gone through phases of rubber extraction, agricultural development, immigration, road building and the promotion of large-scale cattle ranching, timber extraction, and mining. The expanding land use activities in Rondônia have resulted in an increase in the human population and the rate of deforestation so that by 1993 about 25% of the forest area had been lost. In Rondônia, the main activities responsible for deforestation are small farmers, cattle ranchers, miners, and loggers. The implications of the Rondônia Natural Resource Management Project (PLANAPLORO), which is just now being implemented, are critical to the future of the environment and economy of the region (Pedlowski et al., 1997).

In the maps above (Fig. 7.1), we compare the land use change in Ariquemes in three main time periods: 1985, 2000, and 2021. From the data collected, we can infer that the use of agricultural and pasture has increased significantly over the last three decades, with soy being the most important temporary crop produced. Subsequently, the forest areas have decreased tenfold. Urban areas, on the other hand, have increased steadily over time and have not been the major driver of land use change. Another notable aspect is the increase in legal and illegal mining in the region (Table 7.1).

Fig. 7.1 Land cover change in Ariquemes from 1985 to 2021. (Source: Thakuria and Özdemir, with data from MapBiomas (2021))

Table 7.1 Land use change area in Ariquemes from 1985 to 2021

Land use	1985 (ha)	2000 (ha)	2021 (ha)
Forest	351,511	234,262	146,423
Agriculture	80,919	196,637	278,478
Pasture	80,919	196,633	271,182
Non-vegetated area	1658	4509	6894
Urbanized area	1424	2177	3416
Mining	234	2333	3478

Source: Thakuria and Özdemir, using data from MapBiomas (2021)

7.3.1.3 Normalized Difference Vegetation Index (NDVI) and Normalized Difference Built-up Index (NDBI)

The NDVI is a major indicator of urban climate. It provides indications of the presence of vegetation and the amount or condition of vegetation on a pixel basis. The NDVI is derived by using the surface reflectance of the red band (band 3 in TM and band 4 in OLI) and the surface reflectance of the near-infrared band (band 4 in TM and band 5 in OLI). The NDVI values range from −1 to 1, with positive values representing vegetated areas and negative values representing non-vegetated areas.

$$\mathbf{NDVI} = (\mathbf{NIR} - \mathbf{RED}) / (\mathbf{NIR} + \mathbf{RED})$$

where NIR = band 4 (for Landsat TM—wavelength 0.76–0.90 μm) and band 5 (for Landsat OLI—wavelength 0.85–0.88 μm) and RED = band 3 (for Landsat TM—wavelength 0.63–0.69 μm) and band 4 (for Landsat OLI—wavelength 0.64–0.67 μm).

The NDBI, on the other hand, is an index for identifying and classifying built-up areas or impervious surfaces. The positive values of the NDBI indicate built-up areas and those close to 0 indicate vegetation, while the negative values represent water bodies. The NDBI is expressed as the following equation:

$$\mathbf{NDBI} = (\mathbf{MIR} - \mathbf{NIR}) / (\mathbf{MIR} + \mathbf{NIR})$$

where MIR = band 5 (for Landsat TM—wavelength 1.55–1.75 μm) and band 6 (for Landsat 8—wavelength 1.57–1.65 μm) and NIR = band 4 (for Landsat TM—wavelength 0.76–0.90 μm) and band 5 (for Landsat OLI—wavelength 0.85–0.88 μm) (Ranagalage et al., 2017).

Figure 7.2 shows that a healthy landscape predominated the entire region of Ariquemes, with high NDVI values indicating dense primary vegetation cover in 1997. The urban region is also reflected within the map; however, it is mostly confined to the center along the edges of the major transportation channels. By 2012, the impact of deforestation for agriculture and urban expansion became evident, with a noticeable decline in NDVI values, especially in areas surrounding the city, to the North-west where lies the mines of Bom Futuro region and toward the East where lies the large-scale farms. The current NDVI calculation reflects that there is no healthy primary vegetation in Ariquemes as of now.

Figure 7.3 shows that in 1997, the NDBI values were relatively low, consistent with limited built-up areas and the prevalence of healthy vegetative cover. Subsequently, in 2012, a substantial escalation in NDBI values was observed, reflective of intensified agricultural expansion. This can be confirmed with the NDVI analysis (Fig. 7.2), which when overlayed tells us that bare soil or agricultural and pasture land have significantly increased over this period and there has been a massive tree cover loss. This trajectory persisted into 2023, with NDBI values reaching even higher levels, signifying the progressive expansion of agriculture. However, it is to be noted that roads, or other infrastructural developments like the built-up of the city has been increasing at a steady pace.

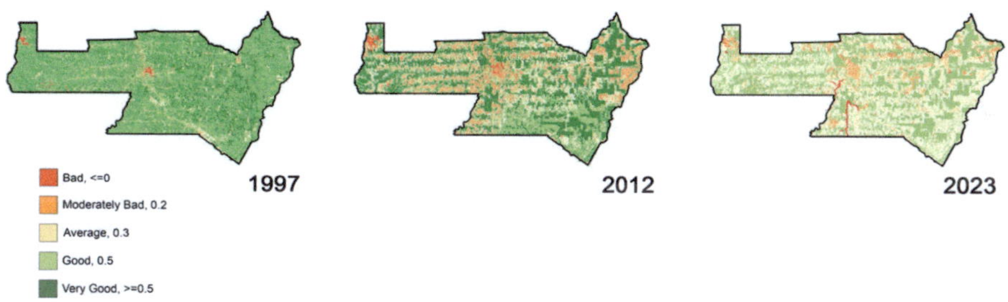

Fig. 7.2 The NDVI calculation for Ariquemes. (Source: Thakuria and Özdemir, based on Landsat images)

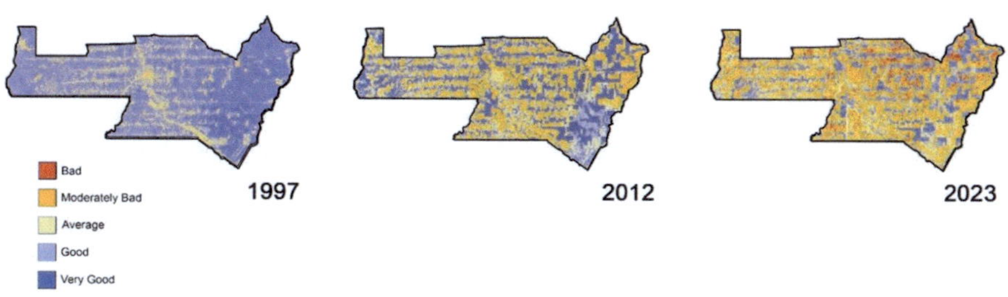

Fig. 7.3 The NDBI calculation for Ariquemes. (Source: Thakuria and Özdemir, based on Landsat images)

7.3.1.4 Topography and Flood Analysis

Ariquemes has a topography characterized by a mixture of low-lying areas and hilly terrain due to its location within the Amazon rainforest region. Specific flood analysis data from MapBiomas (2021) highlights the city's vulnerability to flooding, particularly during the rainy season when rivers and water bodies overflow. The data highlights the importance of robust water management infrastructure and flood mitigation strategies in Ariquemes.

Additionally, it accentuates the critical role of sustainable urban planning in addressing this challenge, incorporating adequate drainage systems and floodplain management to safeguard the city's population and infrastructure from seasonal flooding events. In the city, flooding issues area a major concern for the city's residents and infrastructure. The topographical characteristics, combined with heavy seasonal rainfall, contribute to the recurrent flooding events affecting various parts of the city, emphasizing the need for comprehensive flood risk assessment and mitigation measures.

7.3.1.5 Socio-Economic Analysis

The city's economy is mainly based on cattle ranching and agriculture, which has led to the conversion of forested areas into agricultural land (this phenomenon can be seen in the analysis mentioned above). In addition, the indigenous people historically living in the region have been displaced from their lands due to the urbanization and deforestation of the area. Their former lands are now mostly used for either cattle ranching or agriculture, economic activities that put pressure on the indigenous reserves and threaten the sustainability of the indigenous territories.

Most concerningly, these issues have led to the marginalization of the indigenous people, as the majority of them are no longer able to occupy the lands to which they are rightfully entitled. This leads to problems of social inclusion as well as problems of preserving the heritage and cultural identity of the minorities, as they are no longer able to experience their traditional way of life (Santos & Gomide, 2015).

7.3.2 Micro-Scale Analysis

7.3.2.1 Urban Morphology and Land use

The city is built on a grid layout system adopted on the streets and roads. In addition, the city has a central core in the strategic center of the city. The grid system has immense potential in terms of navigation and land use allocation; however, currently the full potential of this system is not being realized.

When analyzing the morphology of the city, four prominent typologies can be highlighted: The first typology covers the commercial areas; the second covers the rural residential areas; the third is the traditional urban center residential area, characterized by alameda streets, and the fourth covers the contemporary suburban residential areas (Fig. 7.4).

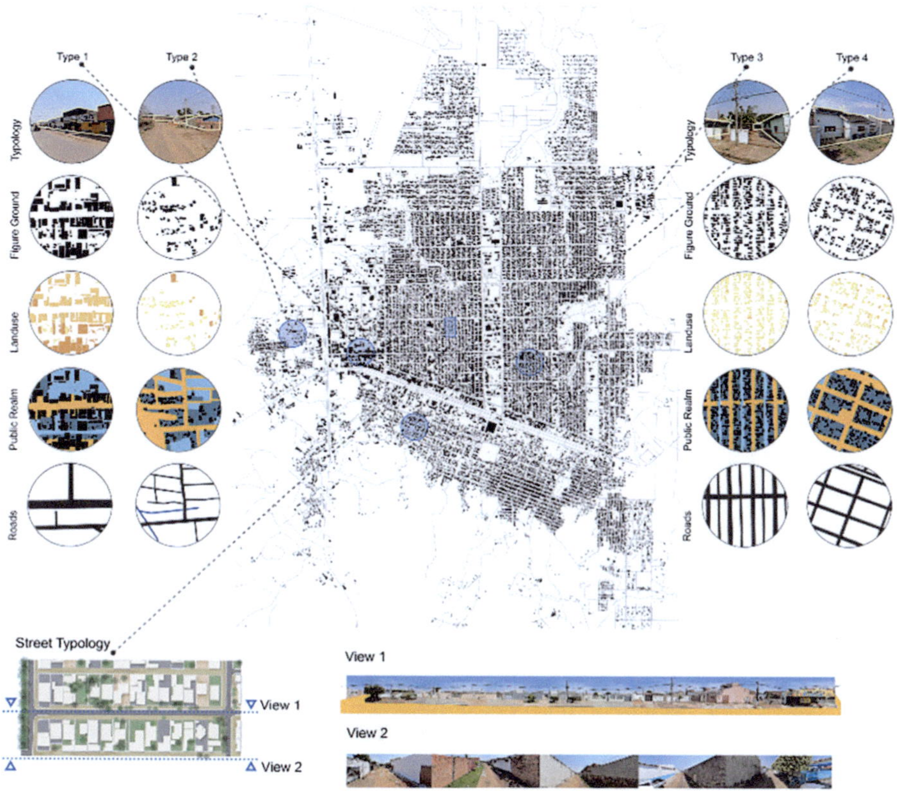

Fig. 7.4 Morphology analysis

There are three main functions, namely residential, commercial and industrial, and institutional (Fig. 7.5). These functions are usually segregated in the form of functional zones. While functional segregation can facilitate focused and efficient resource use, it can also inhibit a holistic and connected urban experience. An approach that focuses on fluid functions and their integration could create a more dynamic urban environment that fosters cross-functional interactions and synergies, contributing to a vibrant, livable city.

Overall, the existing urban fabric is characterized by its density, densely populated structures and very limited open spaces areas. Moreover, the city also lacks green spaces. Due to these issues, the city could be prone to problems of well-being, environmental stability and infrastructural resilience, as well as an unfavorable esthetic appeal of the city.

7.3.2.2 Green and Blue Analysis

Within the urban fabric, there is a notable scarcity of green infrastructure, such as urban parks, green corridors, and natural green spaces. This deficiency is most evident in the denser urban areas. Overall, the green potential of the city has not been

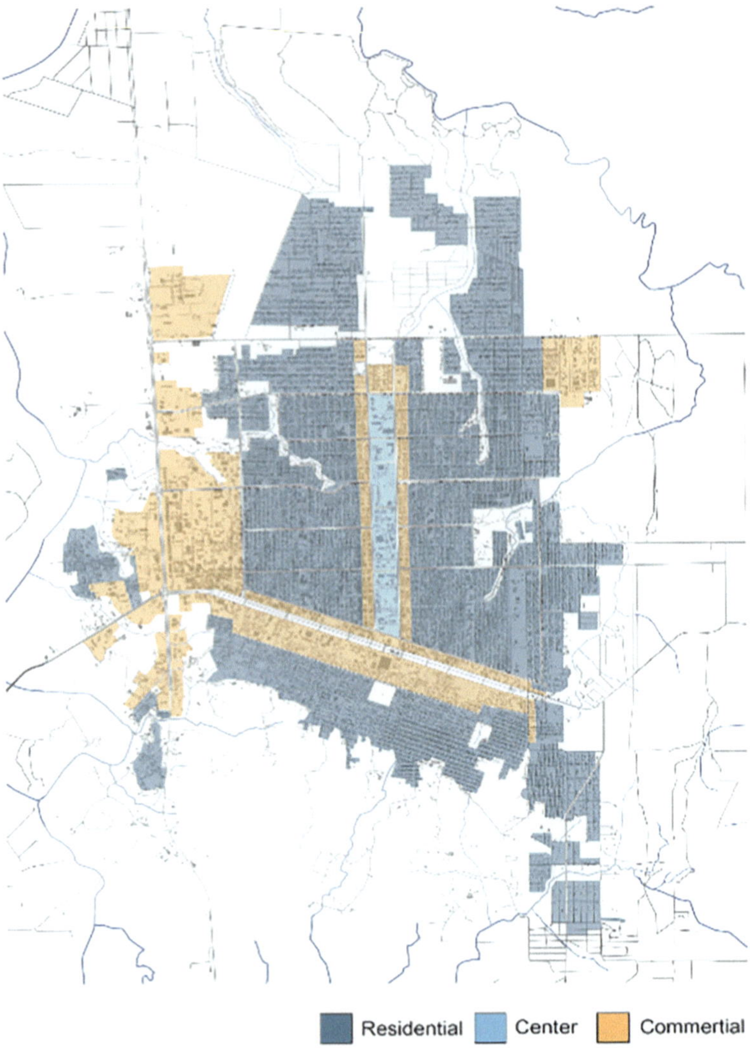

Fig. 7.5 Land use analysis. (Source: Thakuria and Özdemir)

exploited, especially in the central core, where the integration of these elements could enhance the urban landscape and create a sense of connection to the green. It is also important to note that large areas on the periphery of the city are not forest lands, but rather pasture and agricultural land, illustrating the lack of availability of the green space on the city's periphery (Fig. 7.6).

Overall, there is a disconnection between the urban landscape within the city and the landscape present at the periphery and urban edges, which raises the need for a renewed emphasis on green initiatives in order to create a more harmonious and balanced urban environment. It is also important to note that the green function also has the potential to act as a catalyst for a range of functions, including social

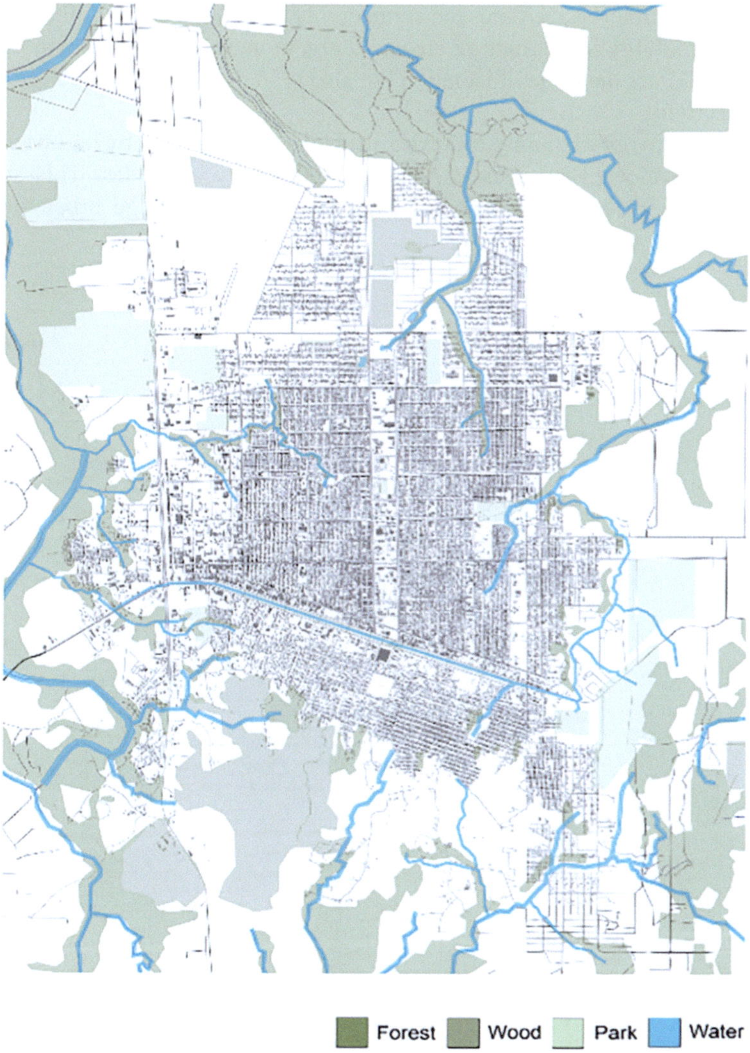

Fig. 7.6 The Green and Blue Analysis of Ariquemes. (Source: Thakuria and Özdemir)

interactions, commercial vitality, and residential improvements. Greenery can trigger a range of positive impacts across different dimensions of urban life.

7.3.2.3 Mobility Analysis

The city is isolated from its surrounding areas, due to inadequate regional mobility infrastructure. Mobility is also an issue at the city level, as public transport options are limited. As a result, people tend to rely on cars for their daily journeys. There is also no rail infrastructure in the city. The lack of a comprehensive railway infrastructure further exacerbates the connectivity challenge. These collective shortcomings in mobility infrastructure create an urgent need for improvement not only to solve to address the issue of isolation but also to improve the overall accessibility and connectivity.

The road network hierarchy follows a structure characterized by highways as the main artery, followed by avenues, streets, and finally alamedas. This structure reflects an order designed to facilitate traffic flow and manage urban functions (Fig. 7.7).

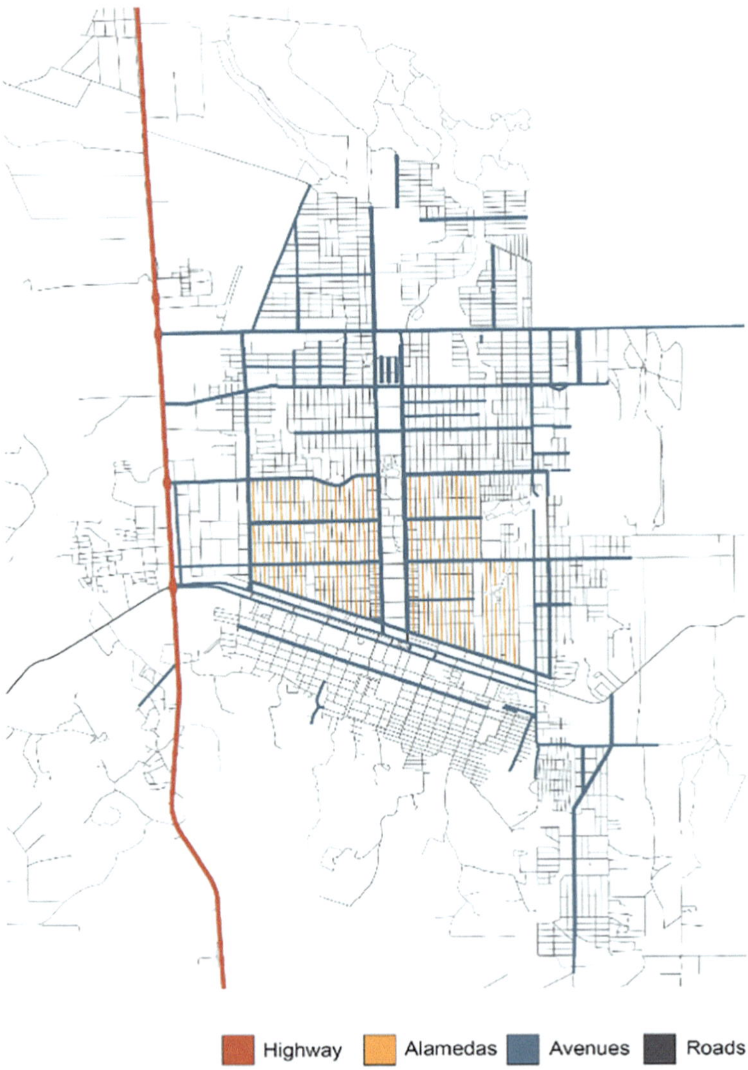

Fig. 7.7 The Mobility Analysis of the Ariquemes. (Source: Thakuria and Özdemir)

Unfortunately, the urban landscape is affected by the deteriorating conditions of both streets and buildings. This collective state of neglect contributes to a less favorable urban environment. This trend extends to the condition of the alamedas, which faces similar challenges and is part of the wider urban degradation.

As the city faces these challenges, there is a clear need for concerted efforts to improve mobility options and infrastructure, as well as the urban landscape. By addressing these issues, the city can work toward fostering a more cohesive and vibrant urban environment that benefits both residents and visitors.

7.4 SWOT Analysis

After the in-depth analysis, we can conclude the strengths, weaknesses, opportunities, and threats observed in Ariquemes, as follows:

7.4.1 Strengths

1. The presence of the city next to several natural resources, such as the Alto Jamari River, several rivers' streams, known as Igarapés, the Natural Reserve Madalena River, among others, proves to be a continuous asset to the environment around the city;
2. Even though the full potential of the grid system is not currently being used in Ariquemes, it has a functioning grid layout;
3. The proximity to larger cities like Porto Velho and the presence of the highway BR-364 prove to be an asset to the regional mobility.

7.4.2 Weaknesses

1. There is no drainage, sewage, water treatment stations or garbage disposal system within the city boundaries;
2. There are greens surrounding the city, however, there is a significant lack of urban greens within the city;
3. The city has no notable public transportation system and even lacks a railway line;
4. Within the grid system, the alamedas, designed as breathing spaces and back alleys for the residential areas, have now been neglected and left in deteriorated conditions;
5. Segregation of land use and building use results in dense and underutilized spaces.

7.4.3 Opportunities

1. There is a great opportunity to integrate the existing forests with agriculture and also introduce agriculture within the city limits to create a more sustainable agricultural environment;
2. There are significant spaces for reforestation combining them with large and vacant pasture lands;
3. The river poses opportunities to create sustainable riparian habitats for the flora and fauna to thrive in.

7.4.4 Threats

1. Monoculture farming is a threat to the climate, the environment, and to the conservation of forests because of its expansion;
2. Due to Ariquemes location in a flat region and significantly close to the Alto Jamari River and its tributaries, it is significantly affected by floods;
3. One of the major problems in this region is the deforestation and lack of greenery in the urban area.

7.5 Problem Definition

The problems in Ariquemes are multifaceted and present several major challenges that require attention and strategic solutions. We can divide the fundamental issues into two categories: physical and socio-economic.

Physical challenges include deforestation, inadequate infrastructure, and unplanned urban growth. Deforestation is driven by agricultural expansion and logging activities. This environmental degradation not only threatens the region's rich biodiversity and contributes to climate change, but also affects indigenous communities and their traditional lands. Inadequate infrastructure, including drainage, sanitation, and water supply systems, exacerbates flooding problems during the rainy season, affecting both urban and rural areas. Unplanned urban growth and deforestation for new settlements promotes the disconnection from the natural environment surrounding the city.

Socio-economic challenges include the segregation of social groups, poverty, and inequality. Addressing segregation requires the integration and empowerment of indigenous peoples, who have historically faced marginalization and displacement as a result of the region's colonization strategies. In addition, issues such as poverty, income inequality, cultural differences, lack of recognition and representation, social inclusion of minorities, and lack of economic diversification needs to be considered. This underlines the importance of comprehensive, sustainable, and culturally sensitive approaches to addressing the issues facing Ariquemes and creating a more resilient and equitable future for the city and its people.

7.6 Vision: 15-Min City Concept, a Grid Planning System

Following the Paris Climate Change Conference in 2015, where the low-carbon city was selected as a policy agenda for large cities, Carlos Moreno introduced the 15-Minute City concept in 2016 as a framework for addressing greenhouse gas emissions (Moreno et al., 2021). The 15-Minute City concept aims to create a self-sufficient neighborhood with the essential functions of living, working, commerce, healthcare, education, and entertainment by decentralizing urban functions and services (Bocca, 2021; Ferrer-Ortiz et al., 2022). Moreno envisioned dense, connected, socially and functionally mixed neighborhoods based on the human-scale design that encourage walking and cycling (Balletto et al., 2021; Noworól et al., 2022).

In our vision for rethinking Ariquemes as a socially inclusive and sustainable city, we propose the 15-Minute grid urban planning concept based on eight dimensions: (1) Proximity to essential services; (2) proximity to public transportation; (3) built density, (4) mixed land use, (5) walkable and cyclable streets, (6) placemaking, (7) inclusivity, and (8) ubiquity. These are briefly discussed below.

1. ***Proximity to essential services:*** Residents should be able to access all essential services such as healthcare, education, work, etc. by foot, bicycle, or other non-motorized vehicle within a reasonable time from their respective place of living.
2. ***Proximity to public transportation:*** The residents should have various public transportation nearby and should be able to access them without any barriers, to reach outside the vicinity of their place of living avoiding using private vehicles.
3. ***Built density:*** The population and employment density of an area support the existence of local businesses and other services. Therefore, residents should be able to access as well as set up businesses or shops within the vicinity of their place of living.
4. ***Mixed land use:*** Residents should be able to access various land uses that fulfill all their daily needs and urban functions close to their place of living. Buildings should be able to cater to a variety of functions, for example, housing, office, and grocery store in one, to create an area that is not only saturated with uses but also creates a safe and vibrant environment for all.
5. ***Walkable and cyclable streets:*** Walking and cycling paths should be well-connected to all the urban functions provided in the 15-minute neighborhood. It should be barrier-free and comfortable for pedestrians, cyclists, and non-motorized vehicle users of all strata. Creating a pedestrian and cycle-friendly

network also encourages small-scale economic development and a more sustainable and eco-friendlier environment.

6. ***Livable public spaces and placemaking:*** Co-creation of spaces together with the people of the community to strengthen the connection and give identity to new spaces according to their needs and uses. This reduces the risk of dilapidated and unsafe spaces in the city.
7. ***Inclusivity:*** All residents should be able to move freely, safely, and without any barriers in public or private spaces, being able to make use of services irrespective of their individual capabilities, age, gender, origin, ethnicity, and so on.
8. ***Ubiquity:*** The above characteristics should be intertwined and widespread so that they are readily and easily available to each and every resident all around the city, irrespective of their socio-economic or demographic status. No space within the city should have any hierarchy.

The 15-minute grid system, as envisioned by us (Fig. 7.8):

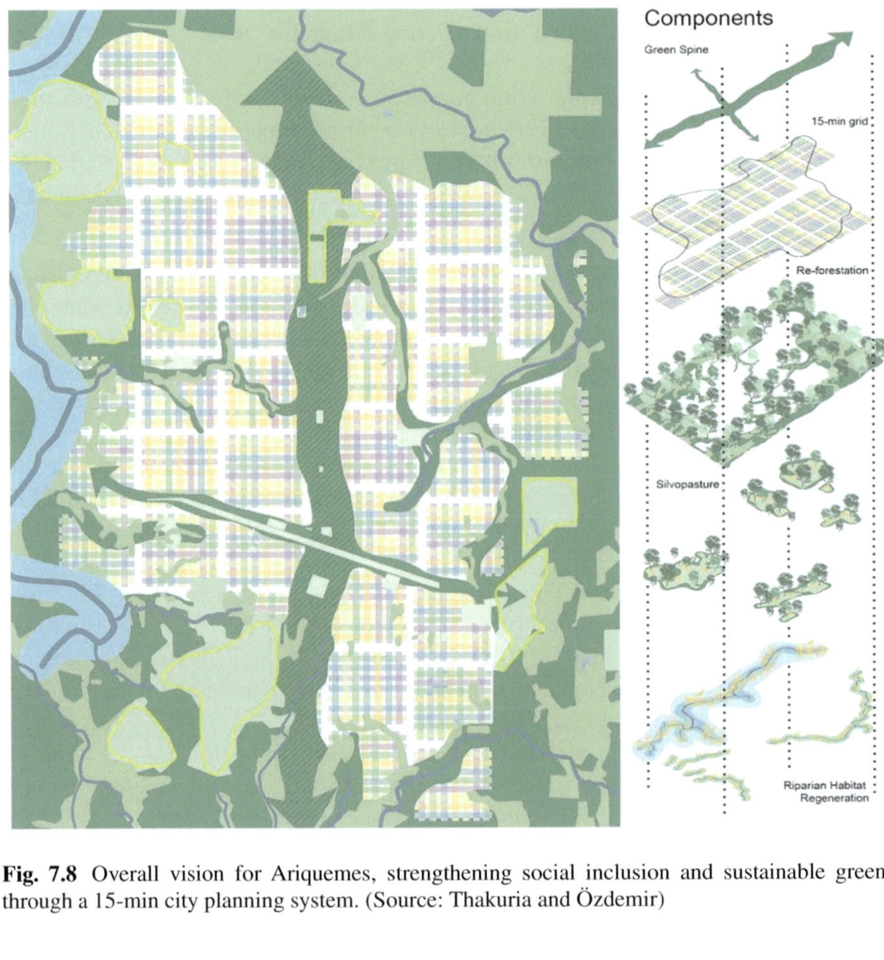

Fig. 7.8 Overall vision for Ariquemes, strengthening social inclusion and sustainable greens through a 15-min city planning system. (Source: Thakuria and Özdemir)

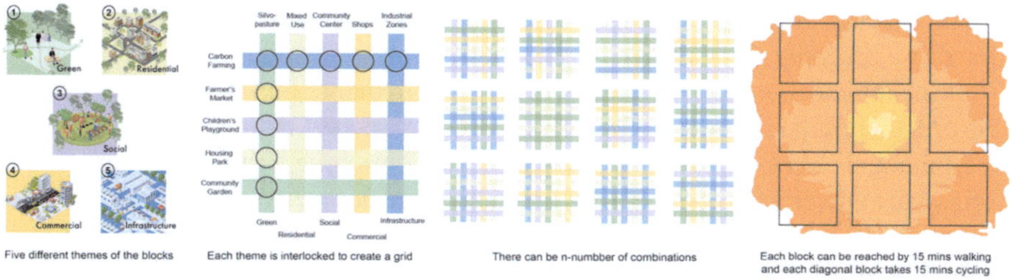

Fig. 7.9 The 15-minute grid system tailored to Ariquemes. (Source: Thakuria and Özdemir)

The "15-Minute" Grid System for Ariquemes is a transformative approach and a powerful tool to address the city's multifaceted challenges while fostering social inclusion, economic growth, and environmental sustainability. This innovative framework seeks to redefine urban living by seamlessly weaving five key themes—social, commercial, residential, infrastructure, and green—into the urban fabric, creating a dynamic and harmonious cityscape. At its core, the 15-Minute Grid System revolves around the idea that every intersection of this weave represents a distinct area with a unique function, contributing to a resilient and interconnected urban ecosystem enriching the lives of its inhabitants, and fostering a more sustainable and inclusive city (Fig. 7.9).

Social Nodes: Within this grid, social nodes are carefully integrated, ensuring that residents are never more than a 15-minute walk or bike ride away from essential community services and cultural amenities. These nodes encompass schools, healthcare centers, community centers and parks, promoting access to education, healthcare, social interaction, and recreational opportunities for all residents.

Commercial Hubs: Sprawling commercial districts thrive at various intersections, nurturing economic vitality while reducing the need for long commutes. Here, local businesses, markets, and retail establishments flourish, fostering entrepreneurship, creating job opportunities, and supporting the local economy.

Residential Enclaves: Amidst the urban weave, thoughtfully planned residential enclaves provide a range of housing options, from single-family homes to high-rise apartments. These areas prioritize green spaces, connectivity, and mixed-income neighborhoods, catering to diverse housing needs while enhancing community cohesion.

Infrastructure Innovation: Infrastructure nodes are strategically placed, addressing critical needs such as efficient transportation, smart utilities, and sustainable drainage systems. Modern transit networks, pedestrian-friendly streets, renewable energy solutions, and comprehensive waste management systems ensure the city's functionality and environmental zones.

Green Avenues: Green spaces are meticulously integrated into the weave, creating area of nature within the city. These areas include riparian habitats, parks, gardens, green corridors and urban agriculture, promoting biodiversity, mitigating urban heat, and providing residents with a connection to the natural world.

The success of the 15-Minute Grid System in Ariquemes lies in its embrace of mixed-use diversity (Fig. 7.10).

At the intersections where themes intertwine, dynamic mixed-use zones emerge (Fig. 7.9). These areas house innovative live-work spaces, cultural institutions, and entertainment venues, fostering creativity, and enriching urban experience. The horizontal and vertical interplay of these themes creates a cityscape where daily life is seamlessly woven into a vibrant spread of functionality. The 15-Minute Grid System not only enhances convenience but also encourages sustainable practices, reducing car dependency, mitigating congestion, and curbing pollution. Our vision for Ariquemes is to promote a harmonious coexistence of urban development and environmental preservation, creating a sustainable and inclusive city for future generations (Fig. 7.8).

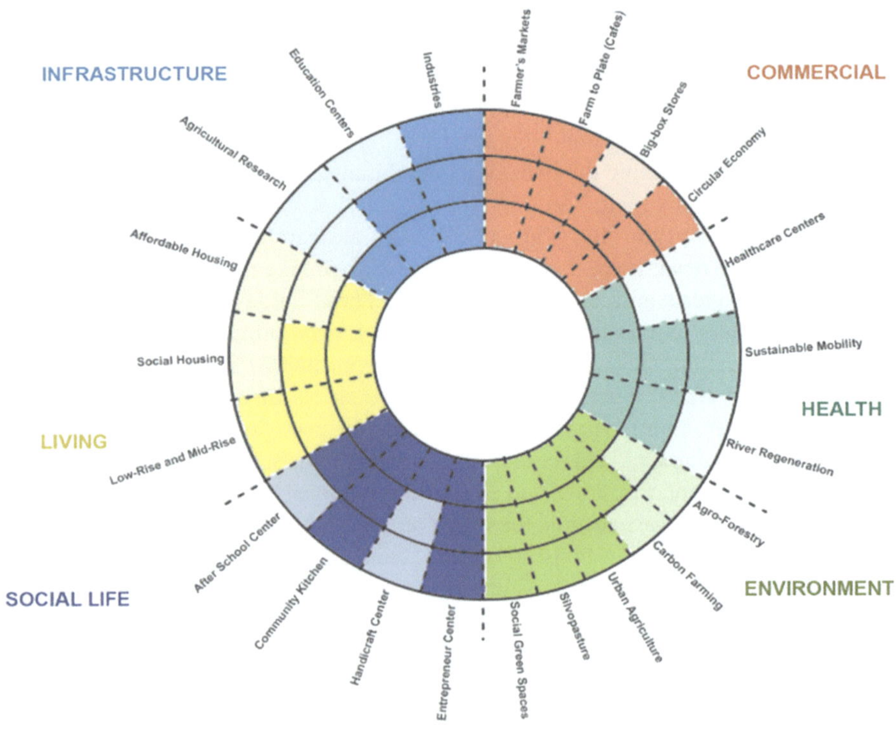

Fig. 7.10 The outcome of the 15-minute grid planning system. (Source: Thakuria and Özdemir)

7.7 Strategic Plan

The strategic plan envisages three main categories of improvements: environmental, infrastructural, and social. These categories are later applied into five distinct layers: infrastructure, commercial, residential, green, and social (Fig. 7.11). To improve the

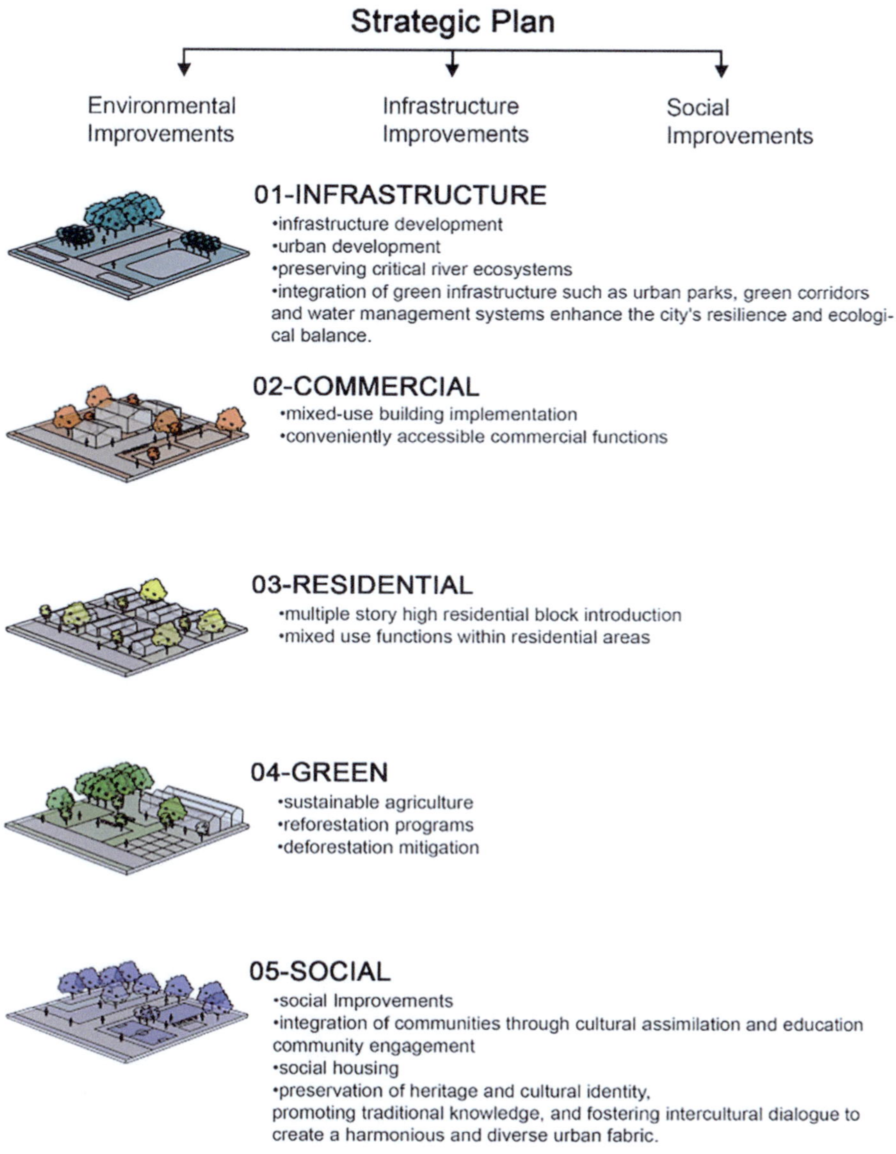

Fig. 7.11 Representation of three improvement strategies and five application layers. (Source: Thakuria and Özdemir)

city as a whole, the plan aims to create an urban landscape that seamlessly integrates a spectrum of improvements, weaving these layers into the urban fabric with renewed vitality. In this way, the city will benefit from a comprehensive transformation that considers environmental sustainability, infrastructure adequacy, social inclusion, and urban well-being.

1. ***Infrastructure***
   (a) Infrastructure development;
   (b) Progressive urban development initiatives;
   (c) Protecting the river ecosystems;
   (d) Incorporation of green infrastructure elements, such as urban parks, green corridors, and water management systems in order to enhance the city's resilience and ecological balance.
2. ***Commercial***
   (a) Implementation of mixed-use building structures;
   (b) Facilitating convenient access to commercial functions.
3. ***Residential***
   (a) Introduction of multi-story residential blocks;
   (b) Introduction of mixed-use functions within residential areas.
4. ***Green***
   (a) Promotion of sustainable agricultural practices;
   (b) Reforestation programs;
   (c) Mitigation of deforestation.
5. ***Social***
   (a) Initiatives for social improvement;
   (b) Integration of communities through cultural assimilation and educational initiatives;
   (c) Providing social housing solutions;
   (d) Preservation of heritage and cultural identity, promotion of traditional knowledge, and encouragement of intercultural dialog in order to create a harmonious and diverse urban fabric.

The strategic plan includes the integration of the 15-Minute Grid concept, which aims to improve the accessibility of daily activities and basic needs within a convenient 15-minute walking or cycling radius from any point of the city. In this case, this concept harmonizes residential, commercial, social, institutional and green functions. A key aspect of this approach is also to improve mobility infrastructure and the reduce car dependency, thereby promoting a healthier lifestyle (Fig. 7.12).

Fig. 7.12 Strategic Map 2050+. (Source: Thakuria and Özdemir)

7.8 Toolbox

The city can be made more vibrant and livable by harmonizing the urban fabric, including green infrastructure and open spaces and by improving environmental balance. The well-being and social inclusion improvements are linked with the better urban experience. In order to achieve this, a methodology is implemented with a complementary toolbox (Fig. 7.13).

Step 1: ***Current condition.***

The current urban condition of the city is characterized by the dense urban fabric in a grid system with lacks functional diversity, forming a monotonous city.

Step 2: ***Building land cover reduction.***

Some of the residential blocks will be removed to make space for the upcoming green infrastructure.

Step 3: ***Reclaiming the green and re-densification of the built environment.***

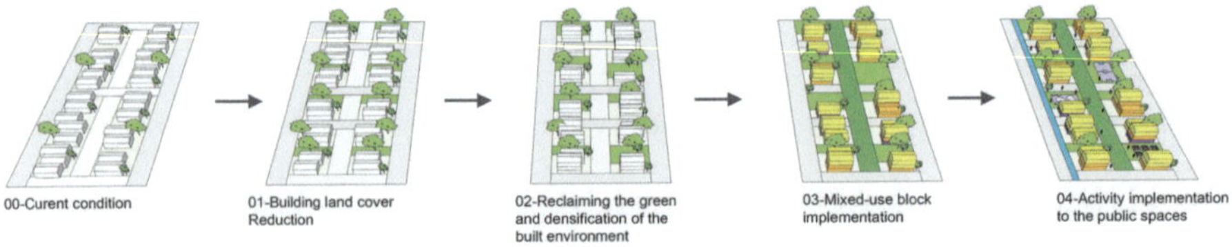

Fig. 7.13 Toolbox, step by step development representation of the selected area. (Source: Thakuria and Özdemir)

Figs. 7.14 and 7.15 View of two redesigned and reimagined alamedas of Ariquemes. (Source: Thakuria and Özdemir)

There will be a green infrastructure implementation to the newly vacant land (Fig. 7.16). Subsequently, some of the buildings will be replaced by new mid-rise blocks.

Step 4: ***Mixed-use block implementation.***

In order to enhance functional diversity, mixed-use functions will be implemented on the new midrise blocks. Moreover, the recreation of alamedas will be undertaken (Figs. 7.14 and 7.15).

Step 5: ***Activity implementation.***

The implementation of social spaces aligned with the new green infrastructure will facilitate the integration of social activities and community engagement along with a recreational corridor (Fig. 7.16).

These proposed steps have the potential to be implemented throughout the city, creating an inclusive framework for the transformation. This collective act aims to reshape the urban fabric for the benefit of a more sustainable, well-connected, and social city (Fig. 7.17).

In the future, as the development of the city continues, the aim is to continue the same strategic plan and concept, preserving the essence of the concept. This approach reflects our commitment to maintaining a consistent course of action, ensuring that future expansions are guided by the same framework that has been set for the urban city's development.

Fig. 7.16 View of an integrated urban farm, with agroforestry and community engagement. (Source: Thakuria and Özdemir)

Fig. 7.17 Masterplan, in the selected zoom-in area. (Source: Thakuria and Özdemir)

7.9 Master Plan

The central core of the original city plan is retained in the new Master Plan (Fig. 7.17), but its full potential as a green spine is now recognized. Therefore, in order to connect the Northern and Southern green proposed in the structural plan, a connective green infrastructure is introduced to the central core. The green infrastructure implementation initiative involves the incorporation of various types of green infrastructure, including botanical gardens, community farms, forest gardens, and so on. In addition, institutional and commercial functions will be implemented along with the green infrastructure, respecting what was already there. These measures will provide the link which was missing in the pre-existing formation of the city, contributing to a holistic urban fabric.

Within the designated prototype area, the strategic plan is integrated with a comprehensive toolbox of measures, characterized by the presence of green spaces and vibrant social areas integrated with a grid system. As the strategic plan covers the 15-Minute Grid concept, accessibility and affinity concepts are also present in the area. As a result, accessible public transport system can be implemented along with the presence of mixed-use functions.

7.10 Timeline

7.10.1 Year 0–5: Social Development

Year 0 Initial Planning: The city government initiates the 15-Minute Grid System project, setting the vision and goals. Public consultations begin to gather input from the community.

Year 1 Social Node Pilots: Pilot social nodes are established in select neighborhoods, integrating schools, healthcare centers, and community spaces. These pilots serve as models for future development.

Year 2 Cultural Hubs: Cultural nodes are designated, including theaters, museums, and community centers. Investments in cultural infrastructure starts, fostering creativity, and community engagement.

Year 3 Green Initiatives: Parks and green spaces are expanded, promoting outdoor activities and enhancing the city's greenery. Sustainable landscaping projects begin to improve the urban environment.

Year 4 Inclusivity Programs: Initiatives to empower marginalized communities, including Indigenous groups, are launched. Cultural preservation and awareness campaigns gain momentum.

Year 5 Education Hub Expansion: The expansion of educational institutions begins, with new schools and universities in strategic locations. The integration of education into the grid system progresses.

7.10.2 Year 5–10: Green Infrastructure Developments

Year 6 Green Corridors: Green corridors are established to connect parks and green spaces, enhancing biodiversity and providing accessible recreational areas.

Year 7 Energy Efficiency: Investments in renewable energy sources, such as solar and wind power, reduce the city's carbon footprint. Energy-efficient infrastructure projects are implemented.

Year 8 Waste Management: Comprehensive waste management systems are introduced, emphasizing recycling and reducing waste generation.

Year 9 Sustainable Mobility: Initiatives to promote cycling and pedestrian-friendly streets are launched, reducing car dependency.

Year 10 Urban Farming: Urban farming projects take place, providing locally sourced produce and enhancing food security. Rooftop gardens and community gardens flourish.

7.10.3 Year 10–15: Functional Developments

Year 11 Mixed-Use Zones: Mixed-use zones at intersections gain prominence, promoting live-work-play environments and boosting economic activity.

Year 12 Community Resilience: Disaster preparedness programs are expanded, including flood-resistant infrastructure and emergency response mechanisms.

Year 13 Affordable Housing: Affordable housing initiatives address housing inequality, providing housing options for diverse income groups.

Year 14 Public Spaces: Public spaces and plazas become focal points for community gatherings, cultural events, and markets.

Year 15 Heritage Preservation: Historic preservation efforts intensify, safeguarding the city's cultural heritage for future generations.

7.10.4 Year 15–50+: Infrastructure and Transportation Developments

Year 15 Transportation Expansion: Comprehensive public transportation systems, including buses and light rail, are expanded. Pedestrian-friendly streets and bike lanes connect the city.

Year 20 Smart Infrastructure: Implementation of smart city technologies enhances efficiency in energy use, traffic management, and public services.

Year 30 Sustainable Water Management: Advanced water drainage systems are implemented, mitigating flooding risks and ensuring a clean water supply.

Year 40 Transportation Innovation: Electric and autonomous vehicles, along with advanced transportation solutions, become commonplace.

Year 50+ Continued Evolution: The 15-Minute Grid System is fully integrated into all neighborhoods, sustaining a vibrant, green, and inclusive city. Ariquemes becomes a global model for sustainable urban living, with ongoing innovation and adaptability at its core.

7.11 Conclusion

This book chapter is based on a comprehensive approach toward reimagining Ariquemes as a sustainable and harmonious city model. It demonstrates the potential for striking a balance between economic development, environmental preservation, and cultural integration.

The implementation of a strategic roadmap embodied in the 15-Minute Grid city planning strategy is a commitment to create a city that thrives in harmony with its natural environment and celebrates its diverse cultural heritage. By integrating green infrastructure, efficient transportation systems, adequate blue infrastructure and cultural preservation initiatives, we envision an Ariquemes that not only mitigates the impact of deforestation but also enhances its future.

The sustainable urban planning strategies were designed to pave the way for a future where improved environmental quality and social well-being go hand in hand with increased economic opportunity. Our vision extends beyond the city limits of Ariquemes; it stands as a testament to the potential for responsible urban development in ecologically sensitive regions. By embracing sustainable green spaces and nurturing the cultural richness of indigenous and minorities communities, we aim to set an example not only for the Amazon, but for cities around the world. In this vision, Ariquemes emerges as a resilient and thriving city that celebrates its unique identity and demonstrates that the harmonious coexistence of nature, culture, and progress is indeed possible in the contemporary urban landscape.

To re-imagine Ariquemes as a model city harmonizing with its environment and celebrating heritage, the strategic planning prioritizes green infrastructure by developing green spaces, urban forests, and parks to enhance biodiversity and mitigate urban heat. Further it tackles sustainable urban planning wherein we implement land use policies that protect natural areas and promote eco-friendly building practices and addressing cultural preservation so as to collaborate with indigenous communities to ensure cultural heritage preservation, establishing cultural centers and organizing events, and so on. The 15-Minute Planning System fosters community connection, making this vision achievable.

Also, balancing agricultural expansion and environmental protection involves, promoting sustainable farming practices, preserving critical natural areas, and nurturing eco-friendly industries, promoting and incentivizing sustainable farming techniques, such as no-till agriculture, crop rotation, and organic farming, educating farmers on responsible pesticide and fertilizer use to minimize soil and water contamination, implementing strict land use regulations to protect critical natural areas, forests, and watersheds from agricultural encroachment, establishing buffer zones around sensitive ecosystems to reduce the impact of agricultural runoff, etc. Through this the project aims to reduce the impact that agricultural expansion has on the ecology and environment of Ariquemes. Furthermore, promoting and investing in eco-friendly industries like eco-tourism, sustainable forestry, and non-timber forest products, encouraging entrepreneurship and innovation to diversify the local economy beyond agriculture, encouraging the adoption of precision agriculture technologies to optimize resource use and reduce waste, supporting research and development of eco-friendly farming technologies, and many more can ensure further agricultural expansion and economic benefits without hindering the environment.

And, to integrate lost Indigenous heritage and culture in Rondônia, the book chapter aims to establish cultural centers, educational programs, and community dialogs to collaborate closely with indigenous communities and minority communities, ensuring their voices are heard in city planning. The goal is to promote cultural awareness and appreciation, foster inclusivity and understanding, participatory planning, and create a safe space for all.

While the 15-Minute City planning system is for the existing grid planning system of Ariquemes, where the vision is to revitalize the already existing, but dilapidated neighborhoods of the city, there always exists certain limitations, with difficulties in implementing changes all over, as the city is constantly expanding. However, in saying that, this concept can always be treated as a prototype with possibilities of executing the idea in newer regions as a part of their urbanistic planning policies and the master plan in the future.

References

Balletto, G., Ladu, M., Milesi, A., & Borruso, G. (2021). A methodological approach on disused public properties in the 15-Minute City perspective. *Sustainability, 13*, 593. https://doi.org/10.3390/su13020593

Bocca, A. (2021). Public space and 15-minute city. *TeMA-Journal of Land Use, Mobility and Environment, 14*(3), 395–410. https://doi.org/10.6093/1970-9870/8062

Cardille, J., & Foley, J. (2003). Agricultural land-use change in Brazilian Amazônia between 1980 and 1995: Evidence from integrated satellite and census data. *Remote Sensing of Environment, 87*(4), 551–562. https://doi.org/10.1016/j.rse.2002.09.001

Carvalho, R. L. S., & Delgado, A. R. S. (2018). Aspects of agricultural activity and socioeconomic development of the municipality of Ariquemes (Rondônia). *Revista Presença Geográfica - Periodicos UNIR, 2*, 2–16.

Paiva, L. N. Q. de and Trevisan, R. (2023) *Ariquemes (RO) e o arq. Antonio Carpintero: projeto e projetista de uma cidade nova em franjas de colonização.* XX ENANPUR, Belém. Retrieved August, 29, 2023 from, http://anpur.org.br/wp-content/uploads/2023/05/st06-09.pdf

Fearnside, P. M. (2005). Deforestation in Brazilian Amazonia: History, rates, and consequences. *Conservation Biology, 19*(3), 680–688. https://www.jstor.org/stable/3591054

Fearnside, P., & Salati, E. (1985). Explosive deforestation in Rondônia, Brazil. *Environmental Conservation, 12*(4), 355–356. https://doi.org/10.1017/S0376892900034482

Ferrer-ortiz, C., Marquet, O., Mojica, L., & Vich, G. (2022). Barcelona under the 15-minute city lens: Mapping the accessibility and proximity potential based on pedestrian travel times. *Smart Cities, 5*(1), 146–161. https://doi.org/10.3390/smartcities5010010

Hänggli, A., Levy, S. A., Armenteras, D., Bovolo, C. I., Brandao, J., Rueda, X., & Garrett, R. D. (2023). A systematic comparison of deforestation drivers and policy effectiveness across the Amazon biome. *Environmental Research Letters, 18*, 073001. https://doi.org/10.1088/1748-9326/acd408

IBGE. (2023a). *Brazilian Institute of Geography and Statistics*. Retrieved August 29, 2023 from, https://ibge.gov.br/cidades

IBGE. (2023b). *Brazilian Institute of Geography and Statistics*. Retrieved August 29, 2023 from https://www.ibge.gov.br/geociencias/cartas-e-mapas/bases-cartograficas-continuas/15759-brasil.html

MapBiomas (2021). *Collection 7*. Retrieved June 2023, from https://plataforma.alerta.mapbiomas.org/

Moreno, C., Allam, Z., Chabaud, D., Gall, C., & Pratlong, F. (2021). Introducing the "15-Minute City": Sustainability, resilience, and place identity in future post-pandemic cities. *Smart Cities, 4*, 93–111. https://doi.org/10.3390/smartcities4010006

Nepstad, D. C., Schwartzman, S., Bamberger, B., Santilli, M., Ray, D., Schlesinger, P., Lefebvre, P., Alencar, A., Prinz, E., Fiske, G., & Rolla, A. (2006). Inhibition of Amazon deforestation and fire by parks and indigenous lands. *Conservation Biology, 20*(1), 65–73. https://doi.org/10.1111/j.1523-1739.2006.00351.x

Nobre, C. A., Sampaio, G., Borma, L. S., Castilla-Rubio, J. C., Silva, J. S., & Cardoso, M. (2016). Land-use and climate change risks in the Amazon and the need of a novel sustainable development paradigm. *Proceedings of the National Academy of Sciences, 113*(39), 10759–10768. https://doi.org/10.1073/pnas.1605516113

Noworól, A., Kopyciński, P., Hałat, P., Salamon, J., & Hołuj, A. (2022). The 15-Minute City—The geographical proximity of Services in Krakow. *Sustainability, 14*, 7103. https://doi.org/10.3390/su14127103

Pedlowski, M. A., Dale, V. H., Matricardi, E. A. T., & da Silva Filho, E. P. (1997). Patterns and impacts of deforestation in Rondônia, Brazil. *Landscape and Urban Planning, 38*(3–4), 149–157. https://doi.org/10.1016/S0169-2046(97)00030-3

Piontekowski, V. J., Matricardi, E. A. T., Pedlowski, M. A., & Fernandes, L. C. (2014). Avaliação do Desmatamento no Estado de Rondônia entre 2001 e 2011. Deforestation assessment in the state of Rondônia between 2001 and 2011. *Floresta Ambient., 21*(3), 1. https://doi.org/10.1590/2179-8087.068213

Ranagalage, M., Estoque, R. C., & Murayama, Y. (2017). An urban Heat Island study of the Colombo metropolitan area, Sri Lanka, based on Landsat data (1997–2017). *International Journal of Geo-Information, 6*, 17. https://doi.org/10.3390/ijgi6070189

Santos, A. M., & Gomide, M. L. C. (2015). Occupation in the environs of indigenous territories in Rondonia, Brazil. *Boletim Goiano de Geografia. Goiânia, 35*(3), 417–436. https://doi.org/10.5216/bgg.v35i3.38834

Tourneau, F. M. L., & Bursztyn, M. (2010). Rural settlements in the Amazon: Contradictions between agrarian policy and environmental policy. *Ambiente & Sociedade, 13*(1), 111–130. https://doi.org/10.1590/S1414-753X2010000100008

WWF. (2023). *World Wildlife Fund*. Retrieved August 28, 2023, from https://www.worldwildlife.org/

Chapter 8
Research on Cacoal Renewal Strategy in the Context of Rondônia Rainforest

Azadeh Ziaee, Xiaoyu Song, and Yuchen Bai

Abstract The Amazon rainforest is a critical ecosystem that plays a significant role in maintaining global climate stability and supporting biodiversity. However, in the Amazonian state of Rondônia, recent urban expansion, agricultural development and rainforest deforestation have exposed topsoil to heavy rainfall, increasing runoff and reducing water infiltration. This has led to soil erosion and the risk of flooding in the urban areas. Given the importance of addressing the water issue in this context, this paper aims to investigate the main causes of deforestation in Rondônia and the city of Cacoal. It also uses integrated analytical methods such as literature reviews, design tools, and case studies to deepen the understanding of Cacoal. By analysing the changing forest cover, altitude, rivers, and transport system, it is shown that the current development of Cacoal faces challenges, risks, and threats. The results look for solutions to these issues, exploring the application of the Sponge City concept in Cacoal. It proposes construction schemes for landscape, roads, vegetation, and buildings, to support the reduction of environmental hazards related to stormwater. In addition, other innovative urban planning technologies can effectively reduce surface runoff and increase water absorption capacity, improving the living standards of residents, and making urban life more sustainable.

Keywords Deforestation · Flood Risks · Sponge City · Cacoal City · Rondônia Rainforest

8.1 Introduction

Of all the world's biomes, rainforests receive the most rainfall. Normally, the rainforest is expected to receive between 2000 and 10,000 millimetres of rain per year (Nasa, 2023). In Porto Velho, the capital of Rondônia, the average annual

A. Ziaee (✉) · X. Song · Y. Bai
RWTH Aachen University, Aachen, Germany
e-mail: azadeh.ziaee@rwth-aachen.de

V. G. Donoso, C. Reicher (eds.), *Rethinking Amazonian Cities in the State of Rondônia*, The Latin American Studies Book Series,
https://doi.org/10.1007/978-3-032-01817-5_8

precipitation is 2.255 millimetres (Weather and Climate, 2023). In the Amazon, rainfall is influenced by different atmospheric circulation dynamics. These are affected by increasing land use changes, global climate change, and deforestation.

Due to the lack of data and the small number of surface meteorological stations installed in the Amazon region, it is still a challenge to understand precipitation variability in this region (Moreira et al., 2022). Despite these challenges, researchers have found that precipitation in the Amazon Basin has changed, with increased rainfall in some areas and more severe floods and droughts, with significant environmental and socio-economic consequences for the region (Cavalcante et al., 2020).

The link between deforestation and rainfall is clear. Climate change in the "deforestation arc" that Rondônia is part of is leading to drier seasonal rainfall. It is predicted that extreme events may become more frequent due to the severity of the reduction in rainfall caused by deforestation, forest fires, and climate change (Mu & Jones, 2022).

Cacoal, the fourth most populous city in Rondônia, faces challenges related to flooding. In February 2022, thousands of families were affected by river flooding following heavy rains. The heavy rainfall caused many rivers to burst their banks, causing flash floods downstream. Vehicles were swept away, and properties were damaged (International Charter Space and Major Disasters, 2022).

This book chapter explores the cause of flooding in Cacoal, Rondônia. It also explores sustainable and ecological tools to address the issue.

From 2002 to 2022, Cacoal lost 21.4 kha of wet primary forest, representing 47% of its total tree cover loss over the same period. During this period, the total area of wet primary forest in Cacoal decreased by 16% (Vizzuality, 2023).

The construction of urban road networks has exacerbated flooding issues, primarily because these roads often penetrate rainforests, disrupting the rainforest's natural integrity and diminishing its inherent capacity for self-repair. Besides, construction is expanding along the roads, taking up more empty land and accelerating deforestation.

Many governments in the Amazon region have taken action to address the loss of forest cover and have recently announced measures to slow down forest loss. Examples include increasing satellite monitoring, stepping up efforts to control illegal logging and slash-and-burn agriculture, and more (República and Brasília, 2004).

To mitigate the damage to vegetation and soil erosion caused by flooding, many previous studies have adopted the model of sustainable agricultural, for example, by adopting conservative practices such as crop rotation, integrated pest management, natural fertilization methods, minimal tillage, and biological control (Gomiero, 2013). However, this is not yet the reality in Rondônia, where agriculture is based on traditional crops (Jackson, 2002; Soule & Piper, 1991; Odum, 1993).

In this chapter, the concept and strategy of the Sponge City is applied in the reality of Cacoal to improve the threat of urban stormwater and make urban development greener and more sustainable. This includes strategies for transforming landscapes and roads, such as creating rain gardens, improving riverbank landscapes, and paving roads with absorbent materials.

This chapter is divided into the following parts: The first part provides basic information about the city of Cacoal, such as its geographical location, population, and historical development. The second part describes the problems faced by Cacoal and analyses the causes and consequences. The third part, the SWOT analysis, is a comprehensive analysis, explaining the advantages and disadvantages of Cacoal in order to eliminate the risk of flooding and discussing the possible development direction in the future. The fourth part presents the specific strategy of Sponge City construction applied in Cacoal. A vision map illustrates blueprints for the future of the Cacoal urban development.

8.2 Methods

In this book chapter, mixed methods are chosen because they allow the triangulation of data, whereby the results of one method can be confirmed or supported by the results of the other. This can increase the credibility and validity of the research. Another advantage is that it can provide a more complete Picture of a research topic.

While qualitative methods can provide a more in-depth understanding of particular issues, quantitative methods can provide a more general overview of a topic. By using both methods, researchers can gain a more comprehensive understanding of the topic. Furthermore, it will outline the best approach for the analysis of these impacts on the Sponge City concept as applied to the main issues of Cacoal city.

This research uses bibliographic research and literature review methods by quoting different publication sources, such as articles and books to learn about the topic of the study and contribute to a deeper understanding of the problem. Case studies were also part of the method to understand the application of the Sponge City concept.

In addition, various tools and techniques in QGIS were used to explore and extract useful insights from spatially referenced information. MapBiomas Collection was used to analyse and map key issues such as land use change, deforestation, urban expansion, and water elements. The MapBiomas Collection (2021) gathers geographic information from satellite imagery from 1985 to 2021. In addition, other information such as NDBI, infrastructure data, topography, and others have been combined.

Many computer software tools such as Adobe Illustrator, Adobe Photoshop, and Adobe InDesign were used to create the graphic figure designs. These tools helped to improve the analysis and design processes. The figures allow readers to understand the solutions created and our vision for the future of Cacoal.

8.3 Discussion

Cacoal is a city in the southern part of the state of Rondônia (Fig. 8.1). It is about 480 km away from the capital, Porto Velho. To the south, Cacoal city borders the municipalities of Rolim de Moura and the Pimenta Bueno. To the east, Espigão D'Oeste; to the north, the state of Mato Grosso; and to the west, the municipalities of Ministro Andreazza, the Presidente Médici and the Castanheiras (IBGE, 2018). The municipal area of Cacoal is 3792.948 km^2 (IBGE, 2010) with a population of 86,895, according to the latest IBGE census in 2022. In addition, it also has a population density of 22.91 people/km^2 (IBGE, 2023).

Cacoal is a young city, founded as a municipality in 1977 (Fig. 8.2). The urbanization of the region began with the construction of the BR-364 Highway. Deforestation increased particularly after 1987, and the 1990s was a period of economic expansion, along with an increase in the number of immigrants, leading to urban and agricultural growth and improvements. Urban development in Cacoal, including the expansion of residential areas, commercial centres, and infrastructure projects began in the 2000s and continued into the 2010s. In addition, the recent expansion of the road network, improvements in public services, and the growth of

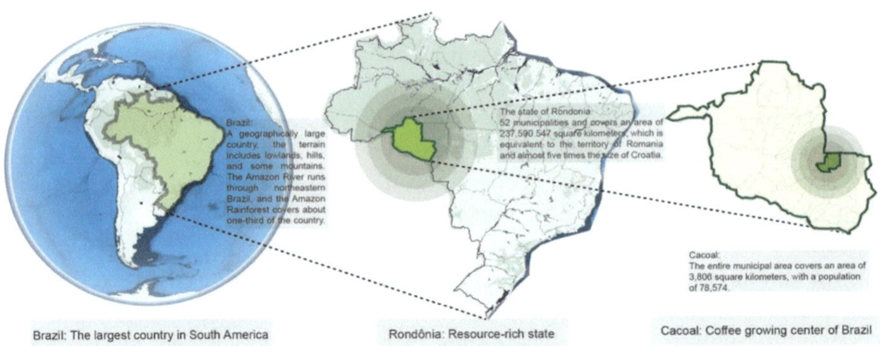

Fig. 8.1 Location of Cacoal-RO. (Source: Ziaee, Song, and Bai)

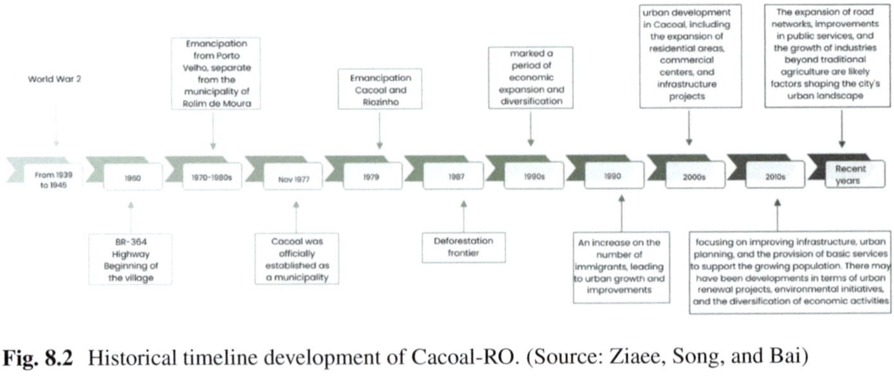

Fig. 8.2 Historical timeline development of Cacoal-RO. (Source: Ziaee, Song, and Bai)

industries beyond traditional agriculture are likely to be factors shaping the city's urban landscape (IBGE, 2023; Cavalcante et al., 2018).

Cacoal is located in the so-called arc of deforestation of the southern Amazon basin. It is important to understand the deforestation patterns, but also the potential of the region, to have valuable information to understand environmental challenges and also conservation efforts in this highly affected area (Michalski et al., 2008; Alvares et al., 2013; Cavalcante et al., 2018).

8.3.1 Problem Analysis

8.3.1.1 Urban Growth

The research on the environment of Cacoal shows that 52.5% of the households have adequate sanitary sewage, and 86.6% of the urban tissue has public roads with trees, as well as 11.7% with public roads with adequate urbanization, with the presence of culverts, sidewalks, paving, and curbs (IBGE, 2018).

Cacoal's industry is growing rapidly, and the demand for agricultural food production has driven the growth of the city's agricultural land. Livestock and crop production have become the main drivers of deforestation in Cacoal. As more and more land is converted to farmland, the rainforest cover is diminishing and the ecosystem is being destroyed.

In addition, the development of Cacoal's urban infrastructure has exacerbated deforestation. As the population grows and urbanizes, so does the demand for housing, commercial areas and transport networks soars. To meet these needs, land across the river has been developed for urban settlements, roads have been paved, infrastructure has been added, and land has been converted to urban use. Among these, the expansion of the road network has played an irreversible negative role in the destruction of the local ecology as will be shown in the next session. Not only has it led directly to the deforestation of large areas of rainforest, but the haphazard expansion of roads has also encroached on basic rainforest and farmland. Biodiversity.

8.3.1.2 Deforestation

Much research in the Amazon region has examined how changes in land use affect the structure of the landscape, taking into account temporal changes during the deforestation process (Armenteras et al., 2006; de Ferraz et al., 2006).

Development of infrastructure in Cacoal city is thought to be directly responsible for 9% to 17% of deforestation in tropical and subtropical countries (Griffin, 2020). The forest remnant of 543 ha has been located within a pasture matrix in municipality of Cacoal city, Rondonia state, Brazil (Cavalcante et al., 2018). This forest fragment represents a region where legal reserves of numerous small- and medium-sized farms converge (Alvares et al., 2013).

Vegetation loss has always been a serious environmental problem, especially for an area as important to the world's ecology as the Amazon rainforest. Cacoal is one of the areas where deforestation is particularly evident, especially the rainforest in the north-central area of city during 1985 to 2023. Figures 8.3 and 8.4 show this evolution. Several deforestation processes can be identified, and three visible steps can be highlighted: First, infrastructure development takes place, with the construction of roads and highways. This leads to forest fragmentation and deforestation. Second, on the axes of the highways, logging, and timber extraction (both legal and illegal) as well as agricultural activities, begin to spread along the axes of the highways, leading to extensive forest loss; and third, land grabbing and land speculation occur when individuals or companies clear land for potential commercial use. Furthermore, land disputes and unplanned land use changes occur, including deforestation.

It is also important to highlight that poverty and lack of opportunity can lead people to engage in activities that contribute to deforestation, such as illegal logging or slash-and-burn agriculture. Inadequate environmental regulations also lead to higher deforestation rates (Michalski et al., 2008; Alvares et al., 2013; Cavalcante et al., 2018; Armenteras et al., 2006; de Ferraz et al., 2006; Michalski et al., 2007).

Our analysis using imagery from Google Earth and MapBiomas database shows the reduction of the green areas from 1985 to 2023. The images show that in 1985,

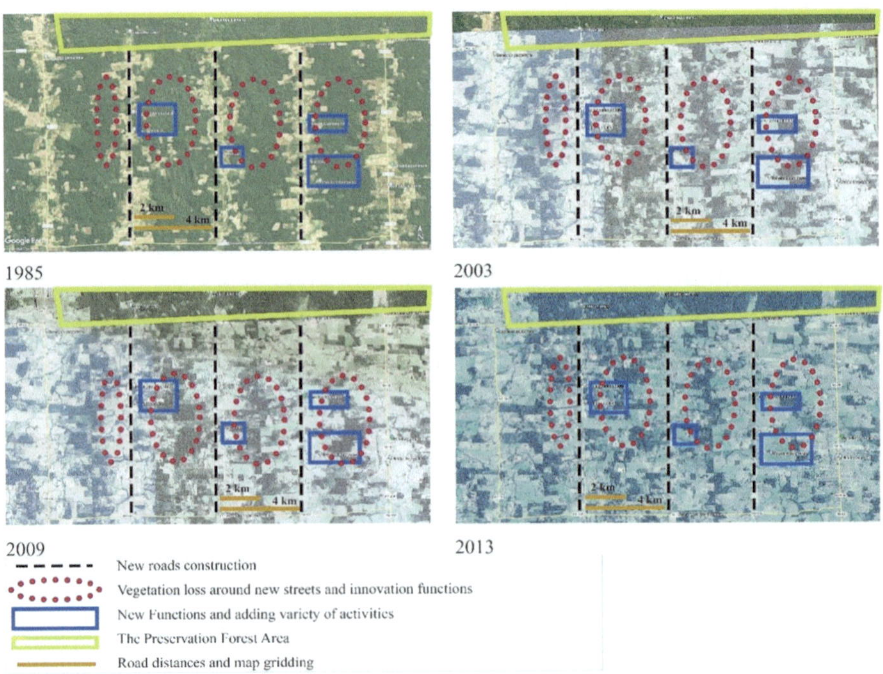

Fig. 8.3 Deforestation in preservation area from 1 to 4 of 1985, 2003, 2009, 2013. (Source: Ziaee, Song, and Bai, based on Google Earth imagery, accessed in 2023)

Fig. 8.4 Deforestation in preservation area from 1 to 42,015, 2018, 2021, 2023. (Source: Ziaee, Song, and Bai, based on MapBiomas data, accessed in 2023)

new roads and infrastructure were built at a distance of 4 km between them, decreasing to 2 km in 2015, which had an impact on the forest. In addition, new buildings and constructions have led to the loss of green areas loss and deforestation in the Cacoal region.

Thus, deforestation and alterations are illustrated in Figs. 8.3 and 8.4.

8.3.1.3 Floods

Between 1991 and 2012 Brazil recorded 4691 flood events, accounting for 12% of all natural disasters in the country. In the Amazon region, the floods affected about 2379 people (CEPED, 2013). However, rivers are a fundamental part of life in the Amazon. Rondônia is located in the southern part of the Amazon region, and it also has a high risk of frequent flooding. As shown in Fig. 8.5, there are six flood-prone areas in the state, one of which is in the Cacoal region.

According to reports from the state government, heavy rains since February 2022 have caused river water levels to rise. This is the case with the Pirarara, Salgadinho, Tamarupá, Riozinho, and Machado rivers. In the municipality of Cacoal, flooding has been frequent, affecting 1400 families. Roads and bridges have also been damaged, isolating many communities (Davies, 2022). Frequent flooding has caused

Fig. 8.5 Vulnerability to Floods in Cacoal-RO. (Source: Ziaee, Song, and Bai, based on data from Snirh, 2024)

damage to the city's infrastructure and disrupted development, and urgent solutions are needed.

The causes of the flooding in Cacoal can be analysed in relation to the extensive deforestation caused by human activities.

Deforestation destroys ecosystems, leading to soil erosion, water quality degradation, and local climate change. Forests play a vital role in sequestering carbon, and their destruction dramatically increases global greenhouse gas emissions, exacerbating climate change. In short, the deforestation of the rainforest for the development of human activities such as urban expansion, agricultural development, road construction, and many more, is the main cause of the frequent rain and flooding in the city of Cacoal. Therefore, it is necessary to carry out specific analysis based on the specific situation of Cacoal, in order to explore more possibilities for the transformation of the city.

8.3.2 *SWOT Analysis*

Located close to the river, Cacoal benefits from an abundance of natural resources, including lush greenery and scenic landscapes. The city's natural beauty is linked together to form a natural green belt of riverside land, demonstrating the unique advantages of being a city in the Amazon region, with enormous potential for ecotourism and outdoor leisure activities.

Cacoal also has a wide range of commercial and educational facilities. For example, there are four universities in the city, which are surrounded by commercial facilities that form an agglomeration. This combination promotes economic development and attracts people from neighbouring areas, contributing to the vibrancy and prosperity of the city.

However, the city of Cacoal also has weaknesses that cannot be ignored. One of the most prominent weaknesses is that the public road transport system needs to be improved. There is also a high demand for private transport, and there are many illegal roads. The pavement has poor water permeability, which hinders mobility and accessibility, and at the same time fails to drain water in the period of the rainy season, causing waterlogging in the city.

In addition, despite Cacoal's proximity to the river, there is a noticeable lack of attractions and entertainment along its banks. At the same time, at the community level, many neighbourhoods lack sufficient public green spaces, which limits the opportunities for outdoor exercise and leisure activities for local residents. There is also a need to improve the streets of Cacoal. In addition to the lack of green spaces, the unmaintained roads detract from the beauty of the city. Finally, Cacoal's public facilities and services need to be strengthened, as these are illustrated in Fig. 8.6.

As mentioned before, Cacoal's location on the river bank makes is vulnerable to flooding. Climate change and poor urban planning can exacerbate this risk, resulting in property damage, disruption of services, and potential threats to human safety. This is also the biggest threat to the development of Cacoal city at the moment. Improving and upgrading public infrastructure requires a lot of resources, and insufficient funds and resources may hinder progress in addressing these infrastructure deficits.

However, this unfavourable situation can be improved in the future by making efficient use of existing resources and opportunities. Good use of the existing rivers, green vegetation and other natural resources, the construction of green gardens, the improvement of the landscape along the streets and the development of the tourism economy are also development paths.

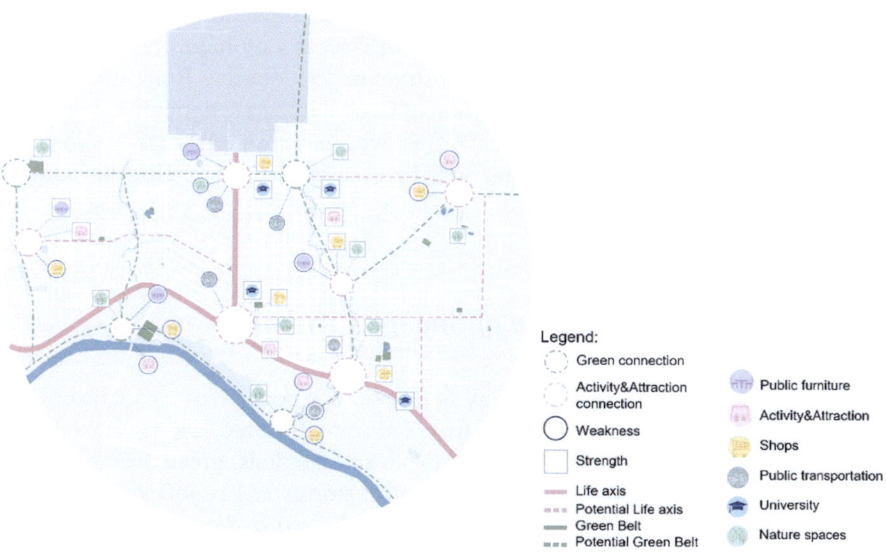

Fig. 8.6 SWOT analysis of Cacoal-RO. (Source: Ziaee, Song and Bai)

The vacant public space in neighbourhoods can develop into community squares, fostering a sense of community and promoting communication among residents. In addition, the local universities can engage academic cooperation, such as setting up a Cacoal Development Research Group to jointly explore the development path of Cacoal.

In conclusion, the city has several advantages, including natural resources, strategic location and educational facilities. However, it also faces challenges such as problems with the road system, underutilized river banks and inadequate public facilities. By leveraging opportunities such as greenfield sites, community spaces, and improved transportation, Cacoal can address its weaknesses and threats to promote sustainable and prosperous development.

8.3.3 Rethink the Urban Development in Cacoal

As mentioned earlier, Cacoal is currently facing the problem of rainforest degradation, which not only leads to a high frequency of urban flooding but also hinders the sustainable development of the city. In this regard, the direction of Cacoal's urban development should be reconsidered. The main challenge is how to make full use of existing resources to create a liveable and sustainable urban environment for its inhabitants, while protecting nature.

Rethinking Cacoal's urban development is an opportunity to transform it into a green, sustainable, live able and friendly city. By protecting natural resources, enhancing green spaces, promoting education and cultural development, and implementing cutting-edge water management technologies, Cacoal can become a resilient, vibrant urban centre that offers residents with an unparalleled quality of life.

As analysed earlier, the two main causes of Cacoal's problems are agricultural development and urban infrastructure construction. So, starting from the problem, can the future urban development direction of Cacoal integrate agricultural development and infrastructure improvement? Can we, as urban planners, realize this potential within the city? This point will be discussed in the following strategy section.

8.3.4 Strategy: Application of Sponge City in Cacoal

The Sponge City construction strategy is proposed. Sponge City is a concept that makes full advantage of the rainwater that is absorbed, stored, and slowly released by urban facilities such as waterways, sunken rain gardens, green spaces, permeable paving, and green roof buildings to control stormwater runoff and allow it to natural accumulate, infiltrate and purify (Li et al., 2017; Xia et al., 2017; Chan et al., 2018).

The application of this concept in Cacoal aims to achieve two overall goals: safety and vitality. By integrating green public space enhancements, it aims to create a safer and more vibrant urban environment for residents.

Before starting planning, there are some principles that need to be noticed. The first point is safety. Like many cities, Cacoal is at risk of urban flooding during periods of heavy rainfall. Traditional urban development with large areas of concrete can lead to rapid surface runoff, overloading flooding drainage systems and increasing the likelihood of flooding. By adopting a sponge city strategy, the city can significantly reduce the risk of flooding and its associated hazards.

The second point is vitality. The Sponge City concept can transform Cacoal into a more active and attractive urban environment. By protecting and enhancing green spaces, cities create a healthier and more enjoyable living environment for their residents. Parks, public gardens, and green corridors serve as natural habitats for wildlife, improve air quality and provide residents with a recreational space close to nature. At the same time, a vibrant urban environment also includes lively public spaces and a strong sense of community. By creating more pedestrian-friendly streets, safe cycle lanes, and convenient public transport options, Cacoal can encourage active mobility and reduce dependence on cars. This promotes social interaction, encourages community engagement, and enhances the overall vitality of the city.

As already mentioned, the application of the Sponge City concept in Cacoal is about safety and vitality. Cacoal can be transformed into a city that not only overcomes water-related challenges, but also provides a high quality of life and a strong sense of community for its residents.

Then, the specific transformation strategy is as follows:

On the landscape side, urban rain gardens are built to increase the city's ability to absorb rainfall. Rain gardens are an integral part of the Sponge City strategy, where stormwater landscape gardens are built in public spaces in the middle of neighbourhoods, with shallow planted depressions that help capture and absorb stormwater so that it infiltrates into the ground rather than causing surface runoff. In addition, native plants can be incorporated into rain gardens to create biodiversity hotspots and support native wildlife and pollinators (Figs. 8.7 and 8.8).

The second point is the restoration of the riverbank landscape (Figs. 8.9 and 8.10). The river is a valuable natural asset of Cacoal, which requires not only protection but also enhancement of the coastal landscape. The creation of a green buffer zone along the river can also act as a natural flood barrier, helping to absorb excess water during periods of heavy rainfall and reducing the risk of river bank erosion. These green spaces can be designed as recreational areas, providing opportunities for residents to engage in outdoor activities and connecting people to the river.

Introducing native plant species into urban landscapes can bring multiple benefits to Sponge Cities. Native plants adapt to the local climate and require less water and maintenance than non-native species. They also support local ecosystems and attract native insects, birds and other wildlife.

Several plants such as *Begonia coccinea*, *Milkania, Xanthosoma*, and so on are not only eco-friendly but also improve the urban heat island effect, where cities tend

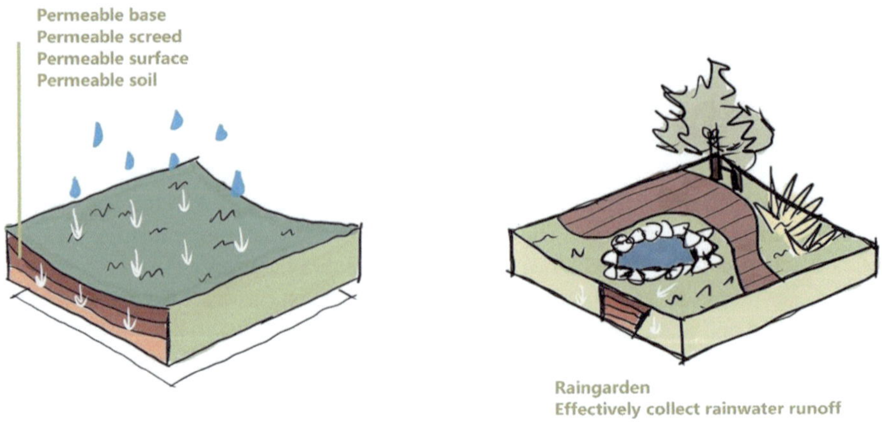

Fig. 8.7 Strategy landscape as Rain Garden. (Source: Ziaee, Song and Bai)

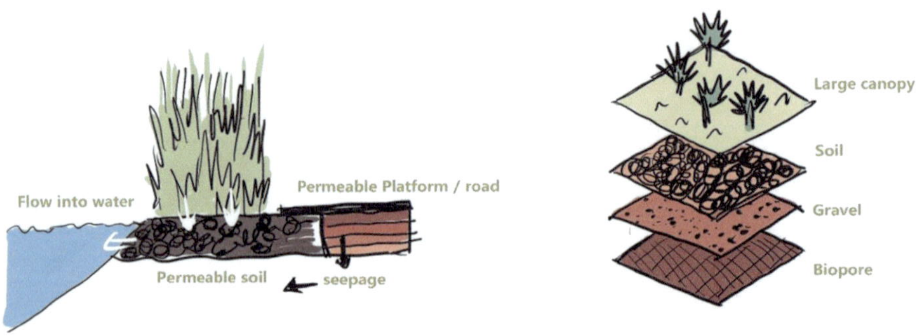

Fig. 8.8 Strategy landscape as Rain Garden. (Source: Ziaee, Song, and Bai)

Fig. 8.9 Strategy landscape as River Bank. (Source: Ziaee, Song, and Bai)

to be significantly warmer than surrounding rural areas due to human activities and abundant concrete. Cooler road surfaces and shaded areas can enhance the overall liveability of the city.

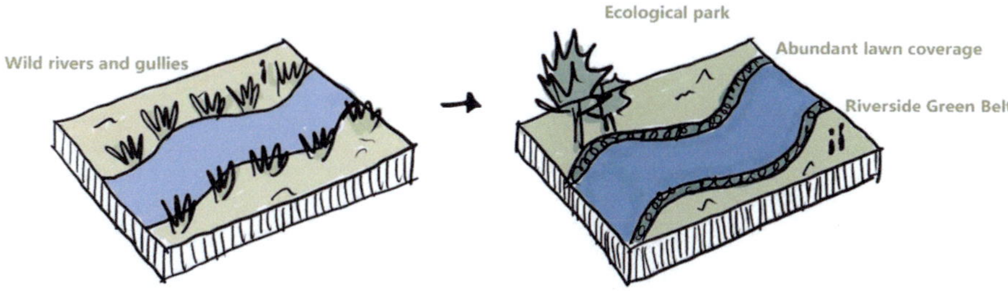

Fig. 8.10 Strategy landscape as River Bank. (Source: Ziaee, Song and Bai)

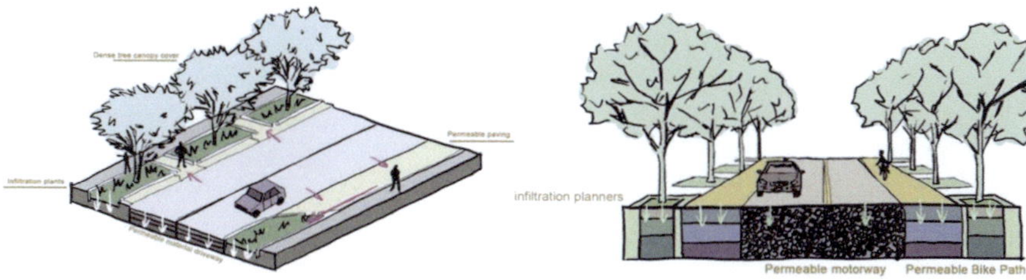

Fig. 8.11 Strategy transportation. (Source: Ziaee, Song, and Bai)

In addition, green building design plays an important role in urban development. Cacoal could incorporate sustainable building practices such as green roofs and green walls to improve building insulation, reduce stormwater runoff, and provide additional green space in an urban environment. This not only helps to improve air quality and temperature regulation, but also creates a more comfortable and healthier living environment for occupants.

In terms of transportation, Cacoal's transport system can play a vital role in the sustainable development of the city (Fig. 8.11). Public transport is encouraged as the first choice of travel mode, and this can reduce air pollution and greenhouse gas emissions. Additionally, the development of pedestrian-friendly streets and safe cycle routes encourage low-carbon travel. Greenery can be planned on both sides of the road to enhance the urban landscape, provide shade, and improve the microclimate. In addition, high permeability materials, such as porous paving, can promote water infiltration and reduce surface runoff.

Finally, a holistic, integrated approach to urban planning is crucial to the successful implementation of a Sponge City strategy in Cacoal, developing a long-term implementation strategy will help Cacoal to carry out urban construction in an open and transparent manner. Overall, it will ensure that landscape and mobility initiatives complement each other, resulting in a more resilient and harmonious urban environment.

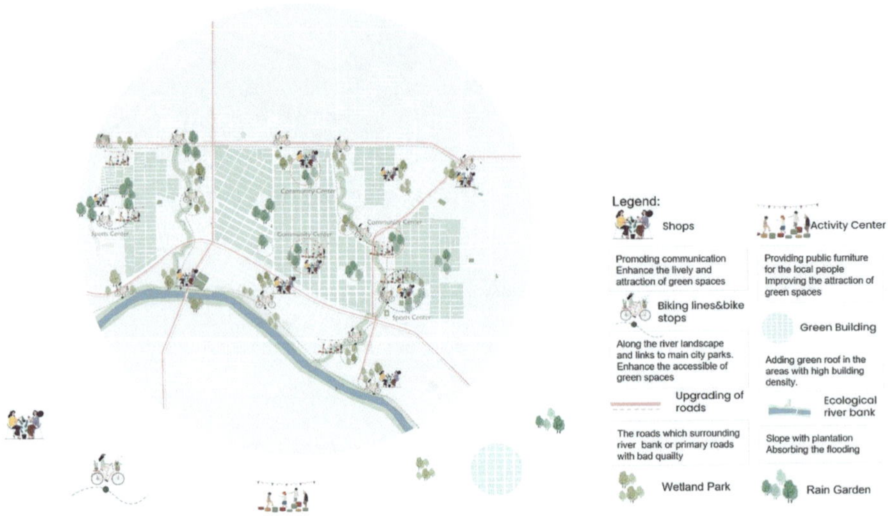

Fig. 8.12 Vision map of Cacoal-RO. (Source: Ziaee, Song, and Bai)

8.4 Conclusion

Although the Sponge City concept has only been proposed for over a decade, many cities around the world have already gained practical experience. In Wuhan, China, for example, more than 300 Sponge City projects have been implemented in the city. By improving rain gardens, roof gardens, and street drainage systems, Wuhan has greatly increased its ability to prevent rain and flood disasters. In addition, many countries in Europe have also implemented relevant practices, such as the ecological corridor in the Netherlands, the green river in Germany, and so on.

Figure 8.12 indicates a vision map to improve the urban development process and rethink Cacoal city.

The rapid development of agriculture in the city of Cacoal, however, ignores the uncontrolled expansion within the city. Although this chapter proposed a strategy for the construction of the Sponge City and suggest specific measures for its application in relation to the actual situation of Cacoal, many challenges remain. For example, funding issues. Cacoal is dominated by agriculture, and limited tax sources make it difficult for Cacoal to develop a Sponge City. At the same time, gaining public support and participation in the planning and implementation of the Sponge City strategy is crucial. Increasing cooperation and connections between residents, local businesses, and other stakeholders is necessary to build a Sponge City.

Therefore, if the Sponge City strategy is to be implemented, it must be integrated into Cacoal's long-term urban development plan. In the short term, small-scale pilot projects can be carried out to test the applicability of the strategy in Cacoal and use this as a demonstration to attract investments. All in all, the construction of a Sponge City will not happen overnight, it will require time and energy to invest in it, and it will need to be adapted to local conditions and updated in a timely manner to make

Cacoal full of vigour and vitality in the process of urban renovation towards a more sustainable approach.

Acknowledgments We would like to express our gratitude to all those who contributed to the successful completion of this book chapters. Their support, guidance, and expertise were invaluable throughout the research and writing process.

First and foremost, we extend our heartfelt thanks to our Professor Verônica Garcia Donoso, whose unwavering support, insightful feedback, and encouragement were instrumental in shaping the direction and quality of this work.

We would like to acknowledge the invaluable assistance of our research team members, Azadeh Ziaee, Xiaoyu Song, Yuchen Bai, students of the master program "Transforming City Regions" of RWTH Aachen University in Germany, who played a crucial role in data collection, analysis, and discussions, without which this book chapters would not have been possible.

This work would not have been possible without the collective efforts of all those mentioned above, and we are truly appreciative of their contributions.

References

Alvares, C. A., Stape, J. L., Sentelhas, P. C., de Moraes, G., Leonardo, J., & Sparovek, G. (2013). Köppen's climate classification map for Brazil. *Meteorologische Zeitschrift (Meteorol. Z.).*, *22*(6), 711–728. https://doi.org/10.1127/0941-2948/2013/0507

Armenteras, D., Rudas, G., Rodriguez, N., Sua, S., & Romero, N. (2006). Patterns and causes of deforestation in the Colombian Amazon. *Ecological Indicators, 6*, 353–368. https://doi.org/10.1016/j.ecolind.2005.03.014

Cavalcante, T., Câmara Gusmão, A., & Francis Ferrari, S. (2018). Unexpected diversity: The potential role of privately-owned Forest remnants in the conservation of the primates of the highly-impacted Rondônia Center of Endemism, Southwestern Brazilian Amazonia. *Neotropical Primates, 24*(2), 82. https://doi.org/10.62015/np.2018.v24.103

Cavalcante, R. B. L., da Silva Ferreira, D. B., Pontes, P. R. M., Tedeschi, R. G., da Costa, C. P. W., & de Souza, E. B. (2020). Evaluation of extreme rainfall indices from CHIRPS precipitation estimates over the Brazilian Amazonia. *Atmospheric Research, 238*, 104879. https://doi.org/10.1016/j.atmosres.2020.104879

CEPED. (2013). *Atlas Brasileiro de Desastres Naturais 1991–2012*. UFSC.

Chan, F. K. S., Griffifiths, J. A., Higgitt, D., Xu, S., Zhu, F., & Tang, Y. T. (2018). Sponge City. In China-a breakthrough of planning and flood risk Management in the Urban Context. *Land Use Policy, 76*, 772–778. https://doi.org/10.1016/j.landusepol.2018.03.005

Davies, R. (2022). Brazil—flooding Rivers displace thousands in Rondônia. 22nd Feb. Retrieved July 2023, from https://floodlist.com/america/brazil-floods-rondonia-february-2022

de Ferraz, S. F. B., Capao, L. M. S. A. C., & Vettorazzi, C. A. (2006). Temporal scale and spatial resolution effects on Amazon forest fragmentation assessment in Rondonia. *International Journal of Remote Sensing, 27*, 459–472. https://doi.org/10.1080/01431160500259907

Gomiero, T. (2013). Alternative land management strategies and their impact on soil conservation. *Agriculture, 3*, 464–483. https://doi.org/10.3390/agriculture3030464

Google Earth. (2023) *Deforestation in preservation area from 1 to 4 of 1985, 2003, 2009, 2013*. Retrieved July 2023.

Griffin, O. (2020). *Amazon road-building could deforest millions of hectares: Report*. 19 November. Retrieved July 2023 from, https://www.reuters.com/article/us-latam-deforestation-idUSKBN27Z1S6

IBGE. (2010). *Instituto Brasileiro de Geografia e Estatística Censo 2010*. Retrieved July 2023, from http://ibge.gov.br/cidades

IBGE. (2018). *Instituto Brasileiro de Geografia e Estatística*. Perfil dos municípios brasileiros: Saneamento básico, Aspectos gerais da gestão da política de saneamento básico. Coordenação de População e Indicadores Sociais. Retrieved July 2023, from http://ibge.gov.br

IBGE. (2023). *Brazilian Institute of Geography and Statistics*. Retrieved July 2023, from https://cidades.ibge.gov.br/brasil/ro/cacoal/panorama

Jackson, W. (2002). Natural systems agriculture: A truly radical alternative. *Agriculture, Ecosystems & Environment, 88*(2), 111–117. https://doi.org/10.1016/S0167-8809(01)00247-X

Li, H., Ding, L., Ren, M., Li, C., & Wang, H. (2017). Sponge City construction in China: A survey of the challenges and opportunities. *Water, 9*(9), 594. https://doi.org/10.3390/w9090594

MapBiomas. (2021). *Collection 7*. Retrieved July 2023, from https://plataforma.alerta.mapbiomas.org/

MapBiomas. (2023). Retrieved July 2023, from https://plataforma.alerta.mapbiomas.org/

Michalski, F., Nishi, I., & Peres, C. A. (2007). Disturbance-mediated drift in tree functional groups in Amazonian Forest fragments. *Biotropica, 39*(6), 691–701. http://www.jstor.org/stable/30043291

Michalski, F., Peres, C. A., Lake, I. R., & Luiselli, L. (2008). Deforestation dynamics in a fragmented region of southern Amazonia: Evaluation and future scenarios. *Environmental Conservation, 35*(2), 93–103. https://doi.org/10.1017/S0376892908004864

Moreira, R. M., dos Santos, B. C., Sanches, R. G., Bourscheidt, V., de Sales, F., Sieber, S., & de Souza, P. H. (2022). Precipitation variability for protected areas of primary Forest and pastureland in southwestern Amazônia. *Climate., 11*(2), 27. https://doi.org/10.3390/cli11020027

Mu, Y., & Jones, C. (2022). An observational analysis of precipitation and deforestation age in the Brazilian legal Amazon. *Atmospheric Research, 271*, 106122. https://doi.org/10.1016/j.atmosres.2022.106122

Nasa. (2023). *Mapping the Amazon*. Retrieved July 2023, from https://earthobservatory.nasa.gov/images/145649/mapping-the-amazon

Odum, E. P. (1993). *Ecology and our endangered life-support, Systems*. Sinauer Publishers.

Snirh. (2024) *Modulo_Geografico Hidrografia*. Retrieved July 2023, from https://portal1.snirh.gov.br/ana/apps/mapviewer/index.htmlin2017

Soule, J., & Piper, J. (1991). *Farming in nature's image: An ecological approach to agriculture*. Island press.

Vizzuality. (2023). *Forest monitoring, land use & deforestation trends*. Global Forest Watch. Retrieved July 2023, from https://www.globalforestwatch.org/

Weather and Climate. (2023). Retrieved July 2023, from https://weather-and-climate.com/average-monthly-Rainfall-Temperature-Sunshine-region-rondonia-br,Brazil/

Xia, J., Zhang, Y., Xiong, L., He, S., Wang, L., & Yu, Z. (2017). Opportunities and challenges of the Sponge City construction related to urban water issues in China. *Science China Earth Sciences, 60*(4), 652–658. https://doi.org/10.1007/s11430-016-0111-8

Part III
Conclusions

Chapter 9
The Importance of Local Participatory Processes in Achieving Sustainable Changes: The Experience of Two Ecological Planning Workshops in Ariquemes, Rondônia

Verônica Garcia Donoso, Joani Paulus Covaleski, and Carolina Salzano Rocha

Abstract Participatory processes are of critical importance to scientists who wish to integrate local knowledge into research and to transfer the results and outcomes to a local level. Participatory processes also represent a well-established methodological tool for decision-making processes. They facilitate the collective reflection and action of diverse groups at the local level. This chapter will illustrate the scientific process of analysis, integration of local knowledge, and transfer of research results through the examination of two co-planning workshops held in Ariquemes in 2023 and 2024. This chapter will commence with an overview of the historical process of urban planning in Ariquemes, a modernist city in the Brazilian Amazon. It will then proceed to discuss the incorporation of an ecological contemporary approach to shift the modernist planning paradigm, an action that was made with local participatory workshops. This chapter will subsequently examine the post-workshop activities, during which the local knowledge integrated the research analysis. The conclusions of this process highlight the main findings, including the challenge of modernist planning, the transfer of research findings, and the new reflections about the specificities of the local scale, which were acquired by the researchers thanks to the local participatory workshops.

Keywords Participatory process · Ariquemes · Ecological planning · Modernist city

---

V. G. Donoso (✉) · C. S. Rocha
Federal University of Santa Maria (UFSM), Cachoeira do Sul, Rio Grande do Sul, Brazil
e-mail: veronica.donoso@ufsm.br

J. P. Covaleski
University Center FAEMA, UNIFAEMA, Ariquemes-RO, Brazil

V. G. Donoso, C. Reicher (eds.), *Rethinking Amazonian Cities in the State of Rondônia*, The Latin American Studies Book Series,
https://doi.org/10.1007/978-3-032-01817-5_9

9.1 Introduction

Participatory processes are an important strategy for the transfer of academic and scientific knowledge to the local level. Here, we present the cases of two participatory workshops that took place in Ariquemes, Rondônia, in 2023 and 2024. They were part of the research activities of the Georg Forster Research Fellowship (Alexander von Humboldt Foundation), granted to the first author from 2022 to 2024. The second and third authors joined the organization and lead the workshop and research activities as members at the local level.

This research project aimed to investigate the specific characteristics of urban expansion and the loss of natural vegetation in the Amazon region. The project focused on the state of Rondônia, employing two distinct analytical approaches at two distinct spatial scales. The first approach employed satellite imagery to understand land use changes at the regional scale. The second approach involved local co-planning workshops, which were used as a tool to contribute to the future local planning process.

This chapter is dedicated to the local scale aspect of the research. It presents the participatory methodology employed in the workshops and the subsequent findings resulting from the in situ visits and activities. To provide context for the reader, data from the history of Brazilian modern urban planning and the specificity of Ariquemes history are also presented.

This chapter is divided into five sections. The first section provides an overview of the historical process of urban planning in Brazil and the subsequent fragility of this heritage. Brazilian modern planning, which is well-known for cities such as Brasília, shaped the way the Brazilian local governments planned cities throughout the country. This planning was often ideologically driven and disconnected from the natural characteristics of the land. In this first section, some information about the history and legacy of Brazilian urban planning will be presented. Additionally, the specific history of Ariquemes is elucidated, encompassing the natural landscape and the imposed urban planning from the 1980s that shaped the city's form and guided the illogical relationship between the urban tissue and the natural surroundings.

The second section will examine the necessity of modifying the contemporary planning mindset, with Ariquemes serving as a case study. The objective is to transition from the modern urban planning paradigm to a contemporary ecological perspective on city planning. This intention was developed at the local scale, in conjunction with two co-planning workshops, that are shown at the third section. The actions and methodology employed in the local participatory workshops are also described.

The fourth section of this chapter presents the post-workshop analysis and reflection process, which includes the development of an ecological analysis of the land characteristics. The fifth section presents a novel concept, designated as Solutions Originating from Nature (SoN). This concept aims to adapt the NbS idea to the Amazonian context, honouring and recovering the knowledge of forest peoples and integrating it into urban solutions. It represents an ongoing process of the researchers.

The conclusions of this experience highlight the key elements that emerged, with a particular focus on the challenges and potential for implementing ecological planning in Ariquemes. The workshops facilitated the transfer of research findings into practical applications and increased the researchers' knowledge about the specificities of the local scale.

9.2 Brazilian Urban Planning History and Participatory Process

Brazilian urban planning has faced significant challenges throughout its history, including political disputes, decisions, and cultural mindsets. Additionally, the country's unique economic and social structures have contributed to the complexities of urban planning. These challenges are in part a legacy of the country's colonial history, but also reflect the intense urban growth period that began in the 1950s, during which the urbanization rate increased from 36.1% to 81.2% over a period of 50 years (IBGE, n.d. Historical data). This was also when many cities from Rondônia state were planned.

The intense urban growth stimulated the discussion of urban planning, particularly from the 1970s onwards. This was also the time in Brazilian history when the country was under a dictatorship (1964–1985). Additionally, this period is related to the final phase of the modern art movement in Brazil (1922–1978).

During the 1960s and 1970s, the discourse on popular participation as an urban planning strategy gained international prominence. Various reflections were published, including Jacobs' work in 1961 (Jacobs, 2011) on the homogenizing principles of modernist urbanism. In 1968, Lefebvre published a work on the right to "habitat," which emerged as a consequence of the post-war housing crisis. Jacobs underscored the importance of critical thinking regarding urban forms, whereas Lefebvre concentrated on the daily activities of the city and the appropriation of space.

In Brazil, the discourse on urbanism during that same period was specially focusing on development, land occupation and expansion (Pereira, 2017). Concurrently, popular participation has been part of Brazilian urban policy since the late 1980s, when significant institutional advancements were made in urban and environmental planning. This period marked the beginning of the country's transition from military dictatorship to democracy. This was marked by the establishment of the country's Constitution in 1988 and the federative decentralization process that strengthened and granted autonomy to local governments (Rolnik, 2009; Donoso, 2017).

Although popular participation in urban public policies in Brazil has occurred since the late 1970s, it was only after the Federal Constitution of 1988 that it was institutionalized as a central element of actions. This was achieved through the implementation of initiatives such as the City Statute (federal law no. 10,257/2001), the Ministry of Cities, and the Council of Cities, which were established shortly after the approval of the constitution (Donoso, 2017; Pereira, 2017).

Brazilian urban planning occurs at different scales of action. The Master Plan (PD) is the primary instrument of urban policy at the municipal scale, governed by the City Statute. Municipalities develop the PD on a local level, encompassing both urban and rural areas. The PD can also include regional environmental processes and water management.

Furthermore, regional planning is conducted through various legal bodies, such as Federal Law 9433 of 1997, which established the National Water Resources Policy and defines the river basin as a territorial unit. However, the different planning scales are not well connected, which can have an adverse effect on the environmental agenda of policies and planning (Peres & de Chiquito, 2012; Faria and Schvarsberg, 2011; Gonçalves et al., 2003).

Currently, urban planning models in Brazilian cities include popular participation. However, effective participation depends on several factors, including process transparency and political will to promote participation. In general terms, there is still a lack of coordination between actions and a low percentage of participation or visibility for participatory processes (Donoso, 2017).

Additionally, urban interventions in Brazilian cities often neglect the environment, with strategies such as the closure of urban rivers and suppression of native vegetation. This comes from a modernist mindset of urban planning. Despite the global discussion about the shortcomings of modernist urbanism, in Brazil this was a dominant line of thinking among urban planners, who applied the theories in city planning and urban legislation. Brazil's capital, Brasília, is a symbol of a modernist city. The urban plan of Lúcio Costa, from the 1950s, implemented ideals of the Atenas Letter and Le Corbusier, such as urban sectors for different activities and axes positioned in relation to monuments. However, this urban plan disregarded most of the natural territory, proposing an unreasonable airplane shape.

The Brasília modernist plan was not friendly for the local people. In fact, 2 years before the inauguration of Brasília in 1960, the first satellite city was established (Taguatinga). A satellite city is a concept that is particularly related with Brasília area. It is a city that is constructed in the vicinity of a metropolis, with a degree of interdependence with the larger urban centre, for example, for job opportunities and services access. This concept is related to the commuter town idea, which is more commonly known concept in other countries.

Over the following years, more satellite cities appeared. This process occurred because the master plan was not designed to accommodate large populations or certain social groups, such as the workers who constructed the capital. Consequently, the workers were compelled to construct their own living areas, given that the capital was planned in the middle of the Cerrado biome and far away from other cities. This led to the creation of new urban areas in the surroundings of the capital master plan. Lúcio Costa's discriminatory attitudes towards social groups are evident in an extract from an interview conducted in Costa, 1928:

All architecture is a question of race, as long as our people are this exotic thing that we see in the streets, our architecture will necessarily be an exotic thing (...)

people with livid faces, who embarrass us everywhere (...). The race being good, the government is good, the architecture will be good.[1]

This history of Brasilia is essential for grasping the mindset of urban planning during that period, where form and sectorization influenced planners to design cities that were not always inclusive of all populations, as in the case of many Amazonian cities disregarding the original populations. Also, despite the fact that modern urban planning was generating severe urban issues in many cities, such as socio-spatial segregation, traffic, and urban violence, it didn't stop the proliferation of more modern planning all over the Brazilian territory. This was described by Jan Gehl as a "Brasília syndrome" or the "Bird Shit Architecture" (Gehl, 2010). The very unreasonable modern planning was still in vogue for long after the modernist period and may still be today, at least in Brazil. For instance, in the 1980s, many modern cities were planned in Rondônia state, including Ariquemes, as a cascade effect of the Brazilian modernist planning fever. Further details on this particular history of Rondônia and Ariquemes will be provided in the following section.

9.2.1 About Rondônia and Ariquemes History

The Treaty of Petrópolis, signed on November 17 (1903) by Bolivia and Brazil, marked the beginning of the de facto occupation of the territory of Rondônia (Morelato & Cardoso, 2020). In exchange for the territory of the state of Acre, Brazil agreed to construct the Madeira-Mamoré railway. The Devil's Railway, as it became known, was constructed to connect Porto Velho to Guarajá-Mirim, two municipalities in Rondônia. The proposed route would avoid the waterfalls on the Madeira River, thereby improving the flow of rubber from both Bolivia and Rondônia (Souza, 2010).

The railway commenced operations in 1912, following the annihilation of many indigenous peoples, along with thousands of workers, during the construction phase. This also marked the beginning of the decline of the first rubber cycle. The railway was managed by private companies in Bolivia until 1930 before becoming the property of the Brazilian state (Souza, 2010).

A significant number of rubber tappers began cultivating agricultural products, and railway stations were transformed into distribution centres for this production. The destination for this food was Porto Velho, a city established by the installation of the North American company Madeira Mamoré Railway, the builders of the railway (Morelato & Cardoso, 2020).

At the same time as the Madeira-Mamoré tracks were being developed, the Brazilian government sent soldiers to the Amazon, under the command of the

[1] Translated by the authors from the original: "Toda arquitetura é uma questão de raça, enquanto nosso povo for essa coisa exótica que vemos pelas ruas, a nossa arquitetura será forçosamente uma coisa exótica (...) gente de caras lívidas, que nos envergonha por toda a parte (...). A raça sendo boa, o governo é bom, será boa a arquitetura."

Marchal Cândido Mariano da Silva Rondon, with the objective of installing telegraph lines for communication between Amazonas and Mato Grosso. The expedition of the so-called Rondon Commission was the government's intention to integrate the North of the country with the rest of the territory (Gomes, 2019). The rationale was that the telegraph would facilitate exploration in remote and uninhabited regions (Maciel, 1998).

The area covered by the commission was transformed in the 1950s into the territory of Rondônia, a tribute to the Marchal (Maciel, 1998). In addition to the essential activity of telegraph lines, the military was also responsible for contact with indigenous peoples, territorial demarcation, and geographic survey (Gomes, 2019). Researchers from the fields of botany, zoology, and geology accompanied the commission, lending it a scientific air and greater respectability. In addition, the construction of telegraph, railways, and roads across the forest prepared the territory for making the land economically productive (Maciel, 1998).

During President Juscelino Kubitschek's administration (1956–1961), there was a strong focus on development. This period saw the creation of a Target Programme for the construction of the BR-29 highway, which is currently known as BR-364. This highway would connect Cuiabá-MT to Rio Branco-AC, passing through Porto Velho-RO (Cavalvante, 2016). The highway was constructed along the route previously used by the Rondon telegraph line, which subsequently became the foundation for the establishment of new cities (de Souza, 2020).

The following occupation and urbanization of the Amazonian territory of Rondônia was largely encouraged by colonization projects led by the military government in the 1970s. Development policies such as official agricultural colonization programmes shaped urbanization, as well as drastic changes in the area (de Souza, 2020). As part of this context, the municipality of Ariquemes was established through the Managed Settlement Projects (PAD's), initiated by the Institute for Agricultural Colonization and Agrarian Reform (INCRA) (De Paiva & Trevisan, 2023).

The city of Ariquemes was established by two PAD projects, Buradeiro (1974) and Marchal Dutra (1978), when a number of settlers relocated to the region to occupy land provided by the government. The city was designed by architect and urban planner Antonio Carpintero, situated to the east of BR-364 and in close proximity to the pre-existing village in the Jamari River valley (De Paiva & Trevisan, 2023). Its geographical coordinates are 9° 54′50" S latitude and 63° 2′38" W longitude (IBGE). Ariquemes was officially inaugurated in 1977.

The municipality's name honours the extinct Arikêmes indigenous group, inhabitants of the Jamari River valley. The extinction of the Arikêmes resulted from several exploitation periods that drastically altered the forest and impacted them (Meira, 2017; Cavalcante, 2015). According to reports from the Rondon Commission, there were 600 indigenous people from this settlement in 1909 on their first passage through the Jamari Valley. By 1913, this number had been reduced to 60 following an attack by rubber explorers (Meira, 2017).

At the time, employees from the Indian Protection and National Worker Location Service (SPILTN) were sent with the objective of alleviating the conflict between indigenous people and independent (and illegal) rubber workers. In 1914, the government established a telegraph station and the "Posto" Rodolpho Miranda in honour of the Arikêmes through SPILTN and the Strategic Telegraph Lines from Mato Grosso to Amazonas (CLTEMA). The indigenous people were then relocated from their territory to the reduced area of "Posto" Rodolpho Miranda, situated on the banks of the Jamari River, which is now known as Vila Ariquemes. In 1950, a new distribution of these lands commenced, resulting in the dispersal of the Arikêmes families once again. In 1984, there were only two Arikêmes families remaining in the municipality of Ariquemes (Meira, 2017).

Despite the effectiveness of the "Postos" in reducing conflicts between indigenous people and extractive workers, the migration of people to the colonies led to an expansion of the exploitation of the territory, this time for agribusiness (Cavalcante, 2015). Furthermore, in addition to the loss of land by the indigenous peoples, the influx of migrants has had an adverse effect on their culture, deteriorating the social and cultural fabric of indigenous communities (Meira, 2017).

Currently, there are no government initiatives in Ariquemes aimed at promoting the indigenous population's history in the region (Cavalcante, 2015).

9.2.2 A Modernist city in the Middle of the Amazon: Ariquemes

As previously discussed, in the 1970s, government policies focused on urban development and the occupation of territory in several Brazilian states, including Rondônia. The construction of BR-364, which traverses the state of Rondônia from north to south, facilitated the colonization of the region (Cordeiro, 2018) and the establishment of numerous cities, strategically positioned along the highway. Consequently, Ariquemes was initially part of the Brazilian agricultural frontier expansion initiative. Its urban development was driven by the planned agricultural destination of official colonization projects for the region, which was subsequently formalized as a municipality (de Martins, 2019).

The city was planned to serve as a colonization nucleus in accordance with requests from the INCRA in the 1970s (De Paiva & Trevisan, 2023). However, Carpintero's modernist planning aimed for it to be a medium-sized city, which would be part of the set of new urban centres that were emerging in Rondônia.

Ariquemes was strategically planned close to a telegraph station, adjacent to the existing Ariquemes Vila and the Jamari River. Its location was protected by neighbouring highways, which served as barriers to urban expansion to the west (BR 364) and south (RO 257). Regulatory axes were implemented to guide urban development, with two central avenues (Fig. 9.1).

The urban planning of Ariquemes defined several avenues as axes for traversing the urban area. These axes run from north to south and include three wide avenues (Av. Juscelino Kubitschek, Av. Tancredo Neves and Av. Candeias). In the east–west direction, another six avenues were designated as axes (Av. Hugo Frey, Av.

Fig. 9.1 Composition diagram of the Ariquemes/RO form. (Source: Created by the Donoso, Covaleski, and Rocha, based on Paiva and Trevisan (2023))

Machadinho, Av. Guaporé, Av. Jaru, Av. Tabapuã and Av. Canaã). Additionally, at the principal entrance of city, there are three diagonal avenues: Av. Jamari, Av. Capitão Silvio, and Av. Rio Branco (Secretaria de Estado do Desenvolvimento Ambiental, 2016).

The two central and parallel avenues, Av. Juscelino Kubitschek and Av. Tancredo Neves, are approximately 2.5 km from BR-364. These avenues form a structuring axis that delimits the city's Institutional Sector (De Paiva & Trevisan, 2023). This sector is home to a number of public buildings, including schools, banks, support centres, cultural centres, and health units. Figure 9.2 illustrates additional elements of the Ariquemes urban plan.

The collector roads that extend perpendicularly to the Institutional Sector also delineate the boundaries of the neighbourhoods, which were originally designated as sectors. These neighbourhoods are home to a variety of businesses and services situated in close proximity to the Institutional Sector.

The organization of the urban plan follows the Brasília modernist inspiration, with the sector identified by numbers. The blocks, with their rectangular configuration, are organized in a regular lot and arranged sequentially (De Paiva & Trevisan, 2023). Figures 9.2 and 9.3 exemplify the city planning.

The city of Ariquemes is situated on a relatively flat terrain, with higher elevation points situated to the south of the urban area (Fig. 9.4). Figure 9.5 illustrates that the city was planned in a higher location, taking into account the Jamari River area. This is significant given the fluctuations in river water volume and the suitability of the area for urban development.

As previously stated, the city of Ariquemes was designed in accordance with the modernist model of urban design, which is characterized by criteria of rationality and technical character. The involvement of civil society in the discussions surrounding proposals for the city was notably absent.

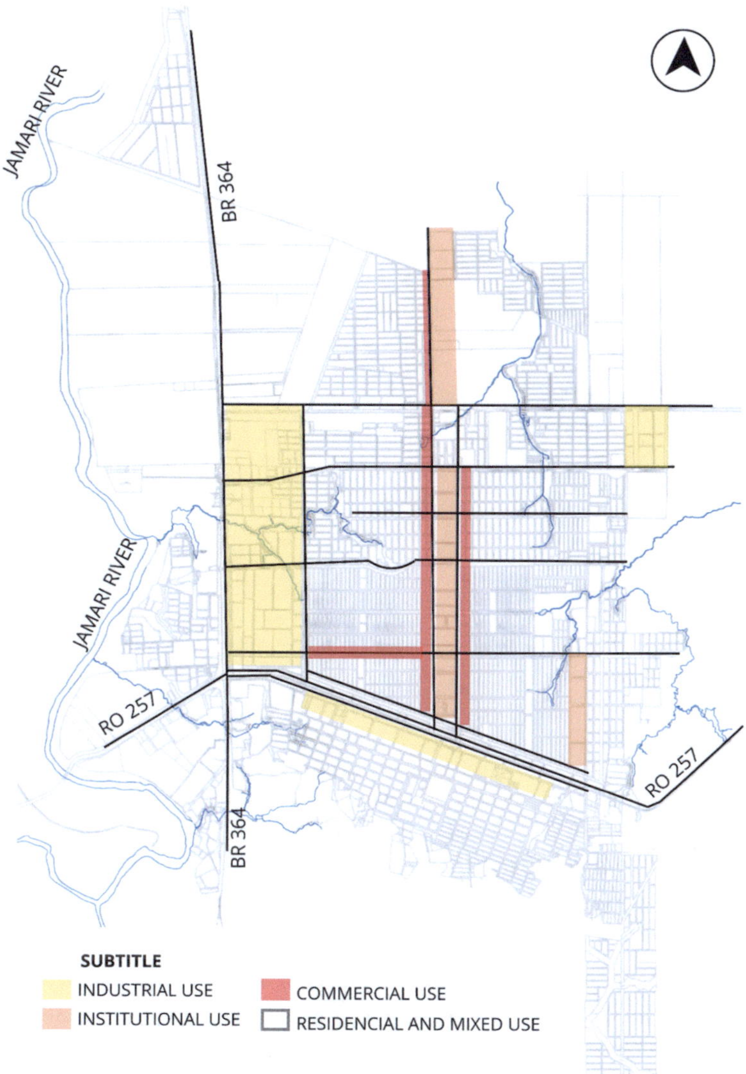

Fig. 9.2 Diagram of the urban zoning of Ariquemes/RO. (Source: Created by Donoso, Covaleski, and Rocha, based on Ariquemes Master Plan (2023))

Fig. 9.3 Drone images from Ariquemes/RO. (Source: Photographs by the authors (Donoso), March 2024)

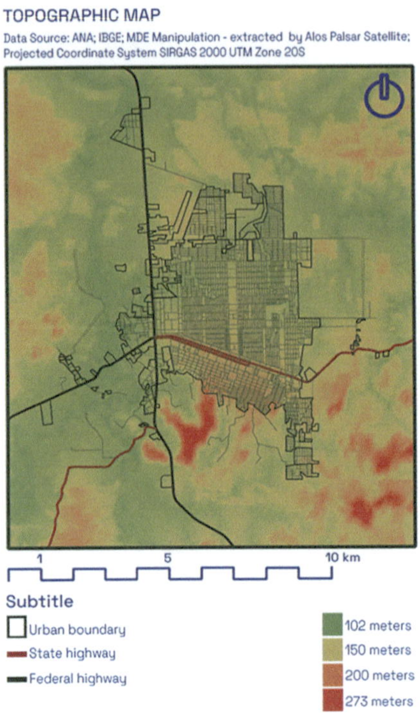

Figs. 9.4 and 9.5 Topographic and Declivity maps. (Source: Authors' own elaboration (Rocha))

9.2.3 Impacts and Fragilities of Ariquemes Modernist Urban Planning

The urban form planned for Ariquemes adheres to a rigid geometric design, regardless of the natural conditions of the terrain. This formal imposition is disconnected from the reality of the Amazon biome. Furthermore, there was a lack of consideration for the distribution of green spaces within the city. Despite the region's rich biodiversity. Ariquemes has 76 neighbourhoods and only 11 open space areas for recreation, such as parks and squares.

Furthermore, the modernist planning brought a rigid sectorization between neighbourhoods, separated in commercial, service, and residential areas, which presents significant challenges to urban mobility. The complexity of circulation makes it difficult to create public transport routes, which currently do not exist. As a result, mobility is mainly dependent on private vehicles, mainly motorcycles, due to their lower cost compared to cars. Bicycles are also used for transport, but they are less common due to the extreme heat, humidity and lack of trees in the public avenues.

The functional and pragmatic approach to urban planning in Ariquemes reinforces a rational and totalitarian reality that is deficient in diversity, affecting the vitality and dynamism of urban life. This approach fosters an environment of individualization and stifles spontaneity. The city that is born under these precepts is segregated, and the zoning establishes a barrier to the possible contribution of the population to the spatial, social, and cultural configuration of the urban environment (Barifouse, 2013).

Some sectors of Ariquemes also received the application of a modernist solution: alleyways (Fig. 9.6). This was the case of Sectors 1, 2, 3 and 4, as shown in Figs. 9.6 and 9.7.

The design of the alleyways was inspired by the superblock's solution in Brasília, where the street hierarchy has a local street where no cars are allowed to circulate. In Ariquemes, the alleyways represent this local street, designed for pedestrians only. The lots in these sectors have two accesses, from the street (frontal part of the lot) and from the alleyway (back part of the lot).

Over time, the alleyways, intended for public use, became a space of abandonment and disinterest for the population, who began to close access to them from the back of the lots. It has become an urban issue and is now a forgotten place with little public life, a place where people cannot walk safely, where vehicles park and where rubbish is dumped (Fig. 9.6).

The region's extreme weather patterns were not adequately considered in the modernist planning process. This is evident in the lack of trees along sidewalks, the limited number of public open spaces, and the high density of buildings on the plots. Additionally, the urban drainage system was not designed to align with the natural drainage basin.

Fig. 9.6 Pictures of the alleyways. (Source: Photographs by the authors (Rocha), March 2024)

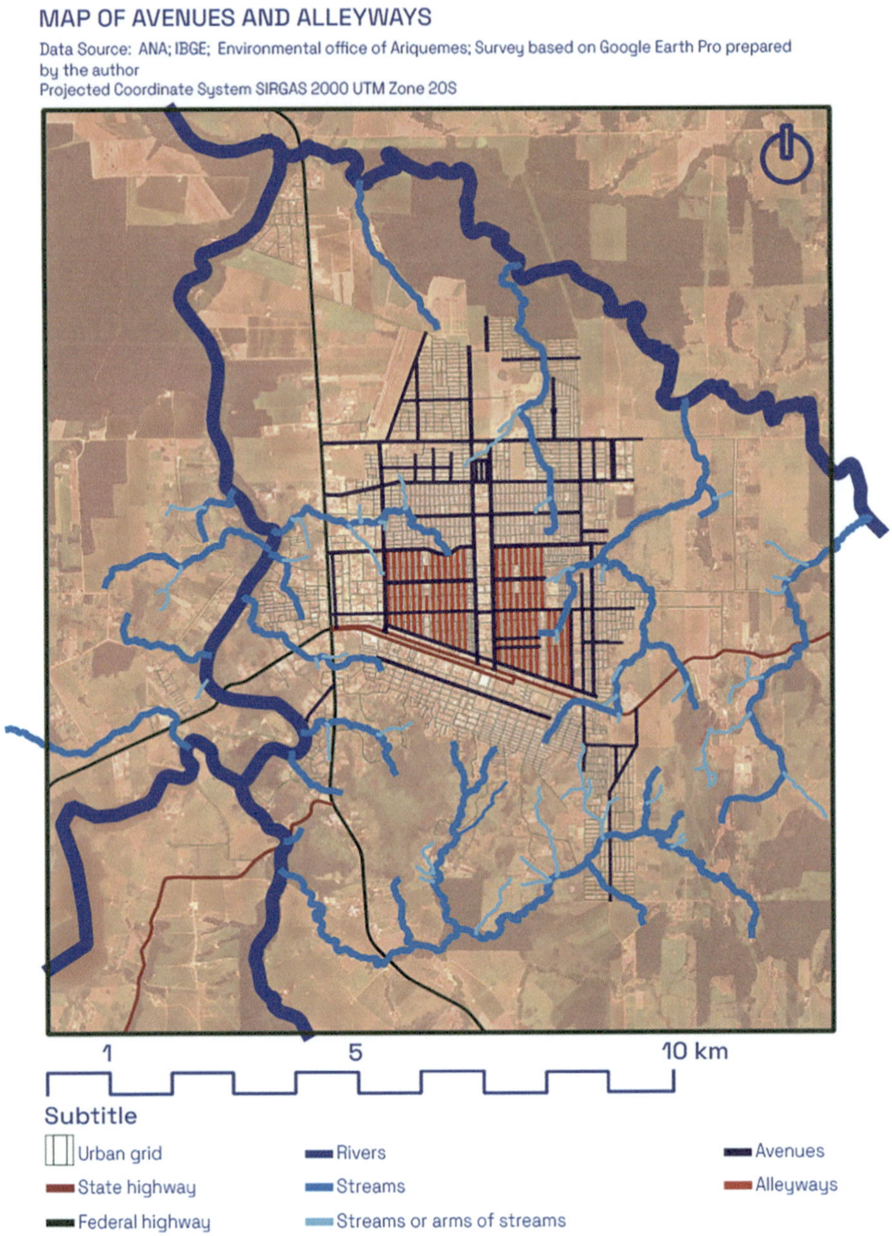

Fig. 9.7 Map of avenues and alleyways in Ariquemes/RO. (Source: Authors' own elaboration (Rocha))

As urban expansion and densification of neighbourhoods have occurred, the incidence of drainage issues has increased (Tucci, 2003). This can be attributed to a rise in the proportion of impermeable surfaces.

Ariquemes is subject to a defined and accentuated rainfall regime over a six-month period (December to May). Together with the urban issues, this results in the occurrence of five common urban flooding points annually, as illustrated in Fig. 9.8.

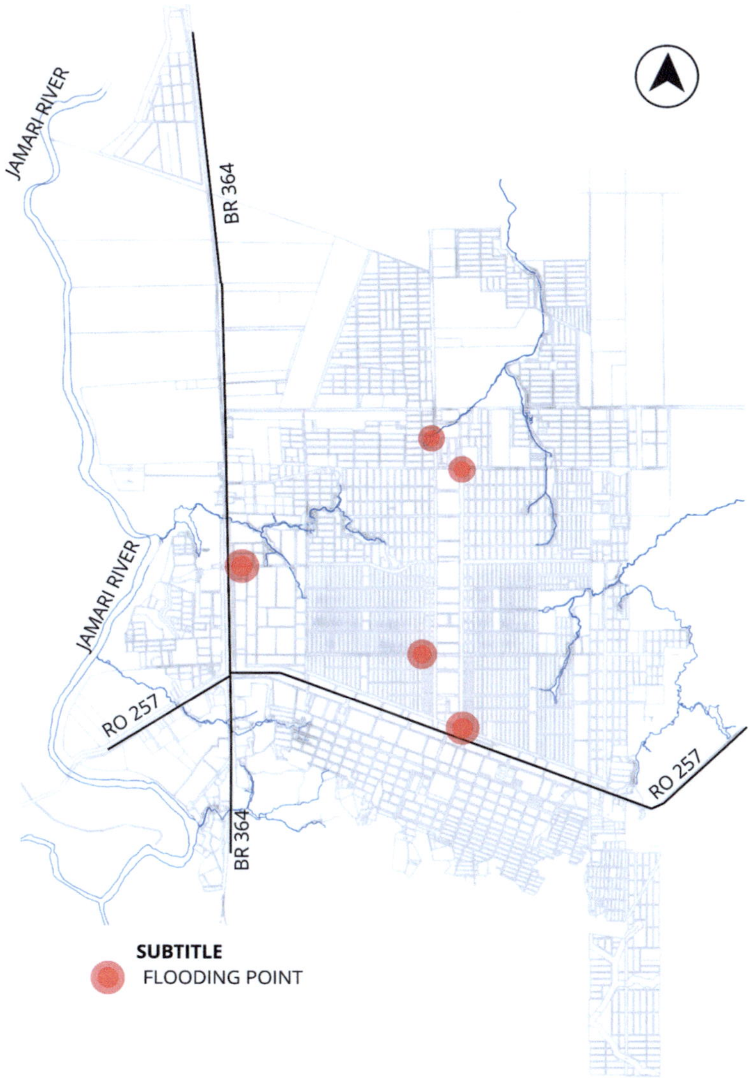

Fig. 9.8 Diagram of the main urban flooding points in Ariquemes/RO. (Source: Authors' own elaboration (Covaleski))

The Water and Sanitation Institute (IAS) has reported that 0.6% of homes in Ariquemes are subject to flooding. Despite the municipality having mapped the risk areas, there are no warning systems in place for hydrological risks (IAS, 2020).

Rapid urbanization and horizontal expansion may have contributed to increased soil impermeability, reducing infiltration capacity and increasing surface runoff during precipitation events. Furthermore, the Ariquemes municipal master plan lacks a definition of permeability index and plot occupation rates, making it challenging to regulate urban permeability conditions.

Another contributing factor to the impermeabilization of the soil is the prevalence of half-land lots in certain sectors of Ariquemes. In Brazil, an urban lot has a minimum area of 125 m^2, with a minimum frontal façade of 5 m. Consequently, the minimum lot area in Brazil is a rectangle with dimensions of 5 m × 25 m (Federal Law 6766/79).

In Ariquemes, the half-lot is defined as the transformation of an urban lot with average dimensions ranging between 10 × 30 and 12 × 2 5 (300 m^2) into two longitudinal lots, approximately 5 m wide and 30 m deep. This is in accordance with the minimum lot allowed by Brazilian federal legislation (Fig. 9.9).

What is particularly worrying about this practice of lot subdivision is that the Municipal Master Plan, in its latest version (Ariquemes, 2019), does not define the occupancy rate, utilization coefficient and/or permeability rate of the lots. These aspects allow an intensive imperviousness of the urban soil, characterized by the absence of environmental limitations in the local legislation.

This practice of excessive imperviousness of urban lots, regardless of their size, is common in many Brazilian cities, such as Limeira (Rossini et al., 2001), São José dos Campos (Gonçalves et al., 2006), and São Paulo (Keppke, 2013). Indeed, there is a cultural practice in the Brazilian private housing construction that includes

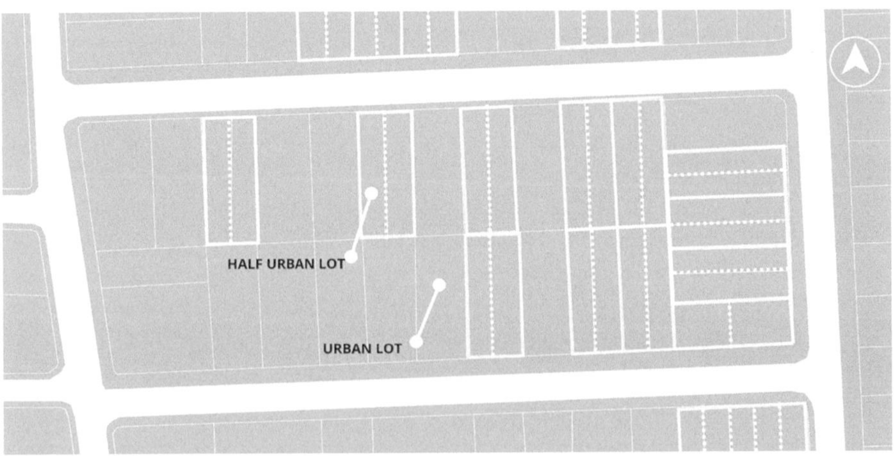

Fig. 9.9 Subdivision diagram of urban lots in the neighbourhood of Ariquemes/RO. (Source: Authors' own elaboration (Covaleski))

impermeabilization of the common areas and reducing the garden to the minimum possible. This practice is justified in many ways, from the expansion of the family nucleus and the necessity of expanding the house area on the lot or even the ease of cleaning impermeable areas when compared to garden maintenance.

In this context, as urbanization intensifies, the incidence of flooding also increases in parallel. This can result in the implementation of inadequate or undersized drainage systems, which are unable to efficiently deal with the volume of water resulting from intense rainfall, a phenomenon that is common in the region, in conjunction with highly impermeable plots.

Furthermore, within the city, approximately 78 streams have undergone alterations to their natural course (Ariquemes, 2023). This has resulted in complications with water infiltration and the environmental quality of river waters due to the presence of illegal waste disposal connections.

9.3 Mindset Change: From the Modernist Urban Planning to the Contemporary Ecological View

The history of planning in Rondônia and Ariquemes comes from imposition. First, the government forced the native indigenous population to move from their land. Second, modernist planning based on formalistic criteria disconnected from the local landscape was imposed. These impositions had a social and environmental cost. A city that does not relate to the landscape and a society that does not relate to its history need to change their mindset to recover these lost links. This can only be done by local groups who are interested in rescuing these memories.

One of the main questions of this research was to ascertain whether the local community was interested in pursuing a change in this situation. Through visits to the city and discussions with various stakeholders, including the local Municipal Environmental Secretary (SEMA) public workers, public authorities, and the scientific community, it became evident that there was a genuine desire to do so. Additionally, it was observed that many local initiatives, spearheaded by SEMA, were already aligned with a contemporary vision of ecological landscape planning. One example is a strategy to mitigate pollution in streams, called the "RG dos Igarapés Project". This project was created by SEMA to bring the community closer to natural spaces and improve their maintenance conditions. This project is one example of the many strategies taking place at the local scale to increase environmental awareness of their natural surroundings.

It is possible to gain valuable insights from these public workers' actions when visiting the city. Those professionals are experienced in raising awareness among the local population about environmental stewardship. This is a time-consuming effort, but a crucial step in achieving sustainable change at the local level. Due to their expertise, they were also key participants at the co-participatory workshops.

With the information in hand about the need to act locally, the interest of the local community, and with the creation of local partnerships, it was possible to plan actions on a local scale. These were created together with SEMA and the local University UNIFAEMA, in order to develop actions that would contribute to what the city and local public authorities needed most. It was in this sense that the workshops were designed.

9.4 The Ecological Co-Planning Workshops

9.4.1 *Process and Methodology*

Participatory processes and methods (PP and PM) are a well-known methodological tool for decision-making processes. It brings together researchers, policy makers, community members, and the general population in the decision-making process to better reflect and act at the local and regional level. Involving different groups and the local population in analysis, planning, decision-making, and all other stages of the project cycle can support a variety of benefits, such as strengthening citizenship, claiming rights, and longer lasting social change.

As mentioned, in the recent years, participatory processes have been used as an important strategy to involve the community in the planning process. This has helped to ensure that planning and projects take into account people's needs and expectations.

During 2023 and 2024, two local participatory workshops were organized in Ariquemes. For it, a mixed method approach was organized, based on the work of QUAPÁ Lab[2] and the participatory workshops organized by the research network on several Brazilian cities since the 2000s, to analyse the open space system and morphological patterns (Macedo et al., 2018).

This multi-method and multi-scale approach was crucial to address the Amazon issue. It combines mapping analysis with local participatory processes and actions. Both are important steps in achieving real changes on the issue of deforestation and reforestation in Amazonian cities. The co-participatory method facilitates the transfer of research results to practical applications and socio-ecological changes. Furthermore, it enables scientists to gain insights at the local level, which is a crucial aspect of the maturation of research ideas.

[2] QUAPÁ Lab is the most important Brazilian research network in the area of Landscape, which includes studies on the Brazilian Open Space Systems and Urban Morphology. QUAPÁ is a network of researchers from 16 Brazilian Universities, coordinated by the supervisory sector of FAUUSP.

9.4.2 The First Co-Planning Workshop, 2023

The first co-planning workshop had two phases: the pre-workshop, from 21/10/2022 to 13/02/2023, and the local workshop, from 27/02/2023 to 03/03/2023.

9.4.2.1 The Pre-Workshop in Germany (2022/2023)

The pre-workshop consisted of a research module course entitled "Ecological Planning for the Amazon Region", offered by Prof. Verônica Garcia Donoso in the Master's Programme "Transforming City Regions" at RWTH Aachen University, Germany. The aim of this module was to give students an overview of the Amazon region—its natural environment, its impacts and its global ecological importance—and to support the ecological planning of the workshop in Rondônia (Fig. 9.10).

The task for this module was to create the material for the workshop, such as posters with maps, analysis, and graphic elements about the regions. A total of 15 students participated in the course, most of them from architecture and urban planning backgrounds. They were divided into three groups and topics: urban, vegetation and water.

The urban group analysed Porto Velho and Ariquemes, highlighting urban growth, land use changes, and morphology. The vegetation group analysed land use changes, deforestation patterns, agricultural and livestock expansion, and forest fires. The water group analysed the river system in the Amazon region of Rondônia, highlighting some of the challenges such as flooding, socioeconomic activities related to the rivers, hydroelectric dams, biodiversity and the impacts of mining and deforestation. The final result was ten posters that were displayed into the local workshop in Rondônia.

The goal was for RWTH students to attend the workshop in Brazil, but due to the lack of economic support, they could not attend the workshop. However, in order to involve them in the event, a presentation video of the RWTH students was shown during the workshop, and a film session about the workshop was organized as an after-workshop activity, also with a presentation and thankful messages from the local students of Ariquemes.

Fig. 9.10 Pictures of the pre-workshop at RWTH Aachen, Germany. (Pictures credits: Donoso (2022 and 2023))

9.4.2.2 The First Co-Planning Workshop in Rondônia (2023)

The 2023 workshop in Rondônia took place from 27/02/2023 to 03/03/2023 with 14 participants (professors, researchers, and students) for the field trip and about 40 participants during the workshop days.

The field trip activities were organized as follows:

- 27 and 28/02: site visit in Porto Velho, specially to the areas of the Museum of the Rondônia Memory, the Cultural Market, the Riverside Park (Madeira-Mamoré historical railway park), Natural Park and Plant Nursery of Porto Velho (Fig. 9.11).
- 28/02 and 01/03: boat trip to the riparian community São Carlos do Jamari, including visit to the community and activities with the local community leaders.
- 01/03: boat trip back to Porto Velho and car travel to Ariquemes.

The field trip was organized with the aim of giving the participants a different understanding of the way people live in Rondônia. The visit to the state capital, the remembrance of the history of the railroad and telegraph lines, and the visit to a riparian community in the middle of the forest were able to provide this overview of different ways of living and interacting with the Amazon biome (Fig. 9.12).

During the boat trip, it was also possible to reflect on the impressive nature of the biome, but also on our human impact, especially due to the large amount of illegal mining that could be seen in many parts of the Madeira River, an activity that

Fig. 9.11 Drone picture of Porto Velho and the historical Madeira-Mamoré railway park. (Source: Photographs by the authors (Donoso), February 2023)

Fig. 9.12 Drone pictures of preserved forest areas in the region of Porto Velho and São Carlos do Jamari. (Source: Photographs by the authors (Donoso), February and March 2023)

heavily pollutes the water due to the use of mercury to separate gold from sand (Fig. 9.13). Some illegal mining dredgers were also seen in São Carlos do Jamari (Fig. 9.14). In addition, deforestation and fires were visible during the boat trip. Some images of the field trip are presented below.

The workshop in Ariquemes was organized as follows:

- 02/03/23 morning and afternoon: site visit in Ariquemes, specially to the Municipal Environmental Secretary (SEMA), Ariquemes Plant Nursery and city visits to squares and alleyways.
- 02/03/23 from 19 h: beginning of the workshop with lectures from: Prof. Verônica Garcia Donoso, Dr. Mayumi Hirye, and the Kanindé Ethnoenvironmental Defense Association.

Fig. 9.13 Pictures from the boat trip between Porto Velho and São Carlos do Jamari. (Deforestation, fires, and illegal mining could be observed. Photographs by the authors (Donoso), February and March 2023)

Fig. 9.14 Drone picture of São Carlos do Jamari and the illegal mining. (Source: Photographs by the authors (Donoso), February 2023)

– 03 and 04/03/23: workshop day with professors, students, and professionals.
– 04/03/23 from 19 h: Presentation of the workshop results.

The lectures opening the workshops had the higher number of participants, around 40. The workshop days of activities had mostly students and professors, in total around 20 participants, due to the lack of availability of some professionals to join the day activities. Even with a small number of participants, the final outcome for the workshop was positive, with SWOT analysis and mapping methods to identify key elements for ecological planning. Some images of the workshop and field trip are presented in Figs. 9.15, 9.16, and 9.17.

The first workshop in Ariquemes was important as a first approach to the local reality. The participatory mapping method and activities allowed the identification of key issues to be addressed, such as drainage problems during heavy rains. This emerged as a major concern for the participants, who observe daily the growth of the city, deforestation, and climate change, with more intense weather events such as increased temperatures and stronger storms. The participants also showed great concern for the future of the alleyways and the need to implement measures to

Fig. 9.15 The first co-planning workshop at the UNIFAEMA University, Ariquemes. (Source: Photographs by the authors (Donoso), March 2023)

Fig. 9.16 Drone pictures of Ariquemes. (The contrast between the historical urban area (Ariquemes village), closed by the Jamari River, and the modern planned city. Source: Photographs by the authors (Donoso), March 2023)

Fig. 9.17 Plant nursery in Ariquemes and Porto Velho. (Cajá fruit (*Spondias mombin*) and Brazilian nut (*Bertholletia excelsa*). Source: Photographs by the authors (Donoso), March 2023)

increase their urban liveability. These results were essential for the planning of the second workshop, specifically designed to address the needs of the local community and to discuss Nature-based Solutions (NbS) as a tool for a greener (and drainage-friendly) Ariquemes.

9.4.3 The Second Co-Planning Workshop, 2024

The second co-planning workshop was organized and created by Prof. Donoso to respond to a local need: professionals from SEMA were interested in learning about Nature-based Solutions in order to implement them as solutions for the Ariquemes drainage system, an action that is also aligned with the actions of the SEMA programme "Greener Ariquemes". The idea was well accepted by the other workshop organizers from the UNIFAEMA University. Therefore, the second workshop was planned as a NbS course.

The 2024 workshop took place from 24/02/2024 to 29/02/2024 with around 55 participants (professionals, professors, researchers, and students) during the workshop days.

9.4.3.1 Pre-Workshop: Volunteer Action

Before the workshop in Ariquemes, a first action occurred. It was a volunteer activity in São Carlos do Jamari. During the inaugural visit to this riparian community, it was possible to engage in discourse with the community leaders, the population, and the district administrator regarding the challenges currently facing the community. Upon contacting the relevant authorities to announce a second visit to Rondônia, it was revealed that a local action could occur at the Municipal Elementary School Henrique Dias. The activity was designed with input from the teaching staff, who requested that the school be painted in order to enhance the learning environment. This voluntary painting initiative was then developed by Verônica Garcia Donoso and Carolina Salzano Rocha in collaboration with the students and teaching staff at the school (Fig. 9.18).

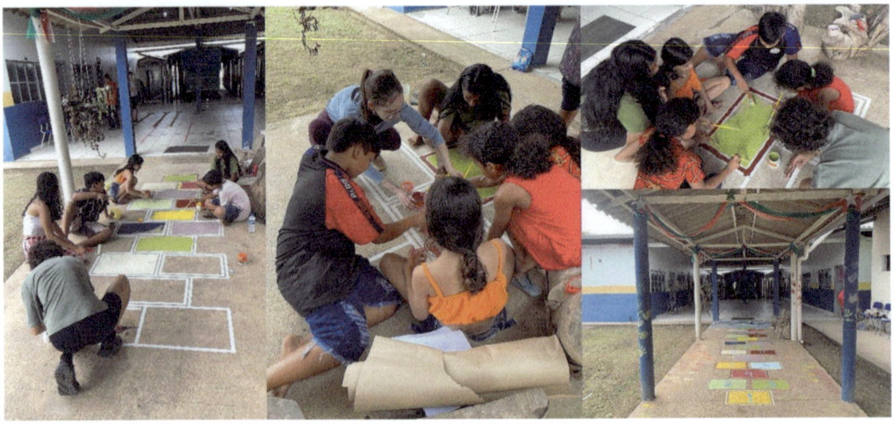

Fig. 9.18 Voluntary work at São Carlos do Jamari. (Source: Photographs by the authors (Donoso and Rocha), February 2024)

After this first voluntary action the co-planning workshop in Ariquemes started.

9.4.3.2 The Second Co-Planning Workshop in Ariquemes (2024)

The workshop in Ariquemes was organized as follows:

- 24 and 25/02/24: Voluntary work at São Carlos do Jamari riparian community
- 26/02/24, afternoon: Presentation of the workshop and class about NbS
- 26/02/24 from 19 h: Master class and activity: SbN—Key concepts for systemic planning
- 27/02/24, afternoon: lecture and activity: NbS examples
- 27/02/24, from 19 h: Co-design workshop for raingardens
- 28/02/24, afternoon: Co-design workshop for raingardens
- 28/02/24, from 19 h: Conclusion of the workshop and presentation of the results
- 29/02/24, early morning: Workshop closing activity: planting native species in a public square.

The workshop was designed to provide participants with information on the NbS concept and types of interventions, calculations of water basins and water capture, selection of plant species for ecosystem services such as phytoremediation and phytofiltration, and examples of solutions from around the world. All course content was delivered using participatory tools to engage participants in building knowledge. In addition, the discussion was specifically aimed at giving them the tools to adapt this European approach and create appropriate solutions for the Amazon biome.

In this second workshop, there were about 55 participants in the opening lectures and about 25 participants in the workshop. They were mainly professionals from the City Hall and SEMA of Ariquemes, Porto Velho and Cacoal, professors from the local universities UNIFAEMA and the Federal University of Rondônia (UNIR), students from both universities and professionals from the city of Ariquemes.

The participatory actions were divided into multidisciplinary and multi-level groups. Based on their diagnosis of drainage problems, each group selected urban areas to develop and design NbS solutions (Fig. 9.19). The sites selected were the avenues of JK, Capitão Sívio and Jamari, the Alleyway of Piquiá and a square in

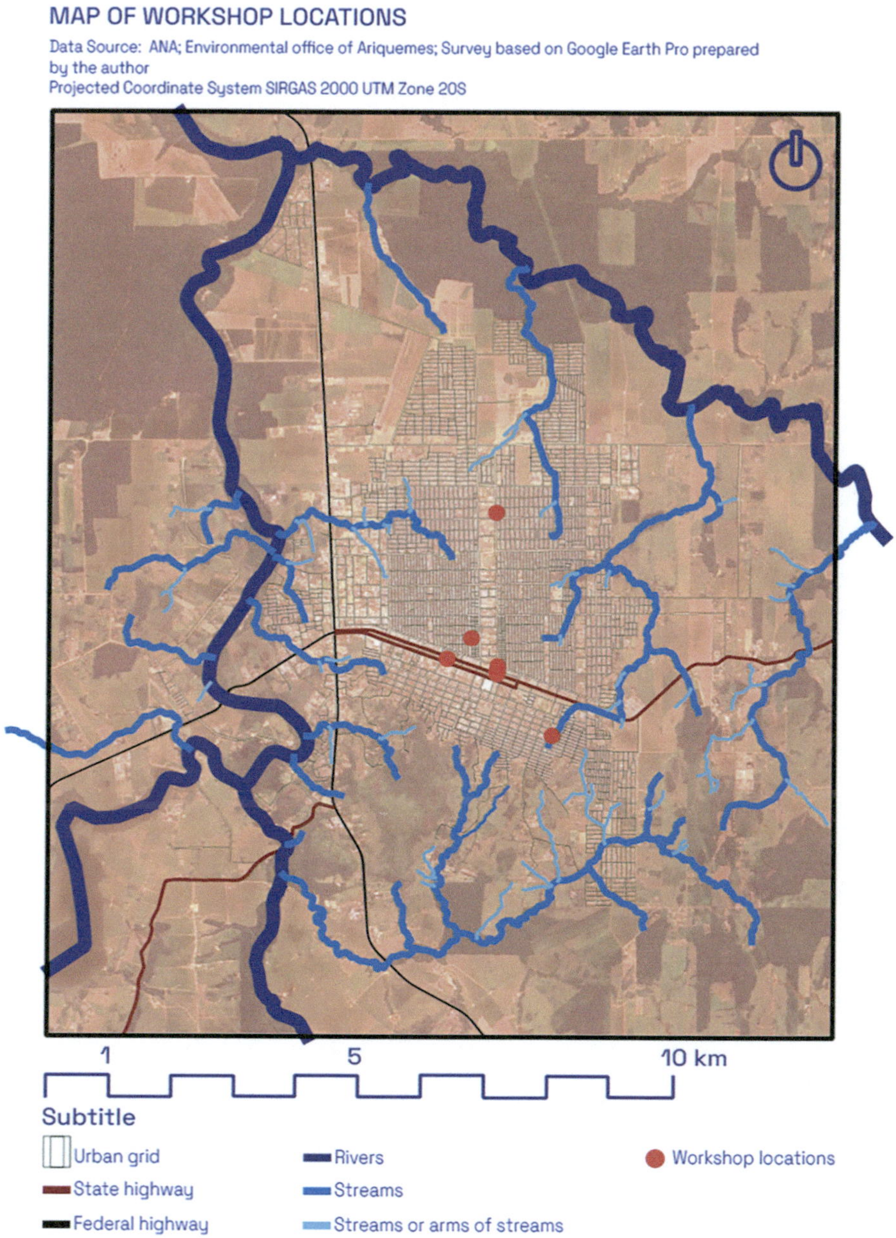

Fig. 9.19 Map of workshop case studies in Ariquemes. (Source: Authors' own elaboration (Rocha), 2024)

Sector 9. All locations were identified by the workshop participants as having inadequate drainage, causing several problems for the residents. The closure of the workshop was with the planting of native species at the "Quero-Quero" square together with the workshop participants and garden crew from SEMA. The outcomes of the workshop included the formulation of further analysis and reflection on the actions undertaken. An evaluation of the challenges and potential of these outcomes will follow later.

The final results of the workshop were published as an unofficial book to help city hall professionals to show the work developed and to stimulate stakeholders and politicians to invest in the solutions. Some pictures of the workshop are presented below (Fig. 9.20).

The two workshops demonstrated the significance of participatory processes and local actions in facilitating tangible changes towards ecological planning and

Fig. 9.20 The second co-planning workshop at the UNIFAEMA University, Ariquemes. (Source: Photographs by the authors (Donoso, Rocha, and Covaleski), March 2024)

environmental awareness. Also, both co-planning workshops initiated a collaborative relationship between institutions, researchers, and public government. With commitment, political, and economic support in decision-making, strategic planning and local participatory actions, this experience and collaboration network can operate supporting ecological and sustainable practices in urban planning.

9.5 Ariquemes Challenges and Potentials for the Future Ecological Planning

This fourth section of this chapter outlines the post-workshop actions. This includes further analysis of the land characteristics as well as reflections on future ecological planning paths and strategies for the city.

The two co-planning workshops held in Ariquemes were of great importance for the researchers. The approach provided insight into the demands of the local level as well as information about previously unidentified local challenges. Additionally, it highlighted the potential areas for the implementation of actions. Further details are provided below.

9.5.1 Challenge: Water Drainage and Water Treatment

Drainage represents a significant challenge in Ariquemes, stemming from a confluence of factors. These include a vast river system, extreme weather patterns, and an urban form that is poorly suited to the topography. However, the most pressing issue has been identified by the local population itself: the urban rainwater drainage system currently covers only 20.4% of the total area, as illustrated on the drainage map (Fig. 9.21).

The Ariquemes Drainage and Stormwater Management Diagnosis Report indicates that the system was proposed in the form of reinforced concrete tubular galleries, measuring 0.60 to 2.00 m in diameter, and square galleries measuring 2.00 to 2.50 m. The objective was to direct rainwater and contribute to the improvement of the city's road system. In this process, surface water is collected from gutters connected to storm drains, and then conveyed via 0.40-m pipes to manholes in the main galleries.

Currently, only 2.17% of the population of Ariquemes is served by sanitary sewage, and only 20.3% is served by the rainwater drainage system (IAS, 2020). The implementation of the New Sanitation Legal Framework (Law No. 14,026/2020) and federal investments to enhance access to basic sanitation throughout Brazil has facilitated the expansion of the sewage network in Ariquemes, that is in 2024 still in construction.

Expanding the basic sanitation network is a matter of public health and socioeconomic development. It is evident that enhancing access to fundamental services,

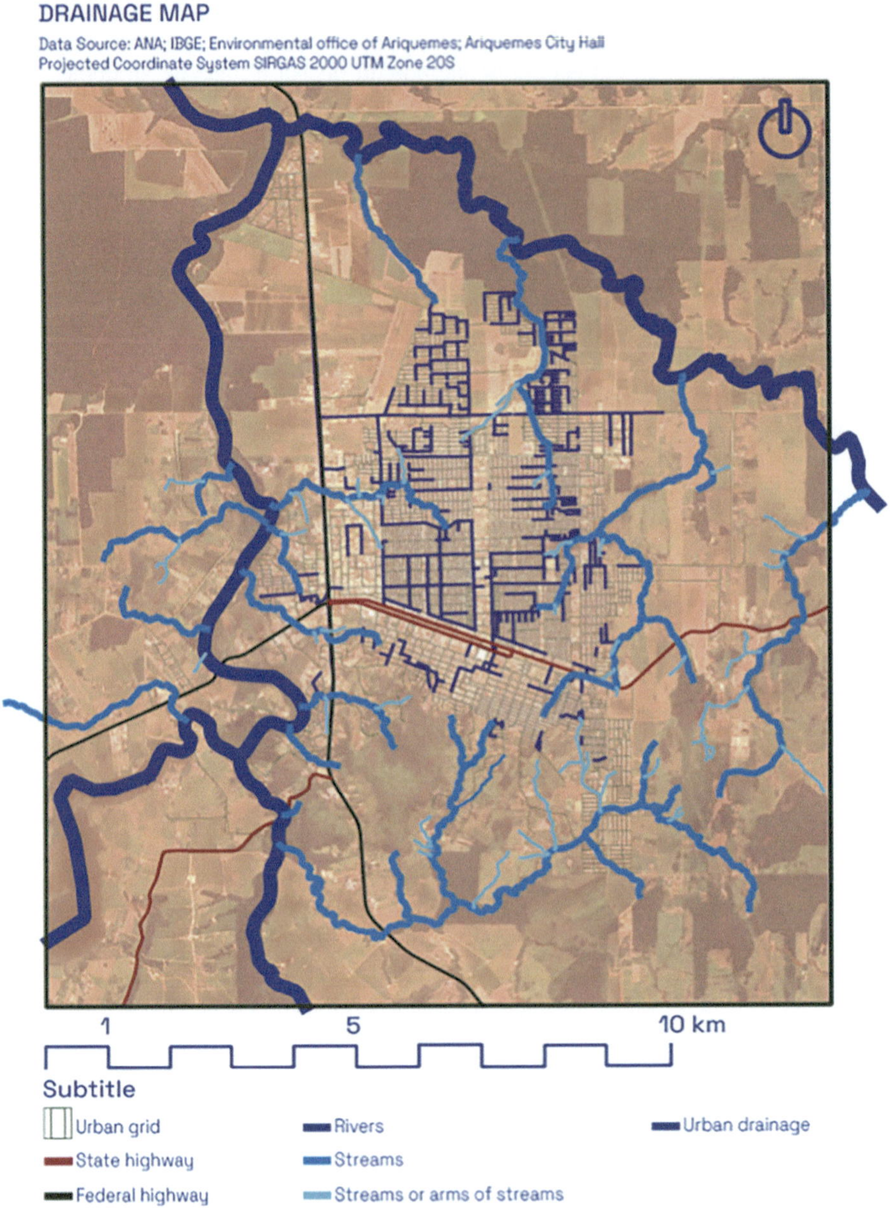

Fig. 9.21 Drainage map. (Source: Authors' own elaboration (Rocha))

such as sewage and rainwater drainage, can markedly enhance the quality of life for residents, in addition to fostering a healthier and safer environment for local economic growth (Carcará et al., 2019).

However, another aspect to be considered is how to perform the expansion of the basic sanitation network, which should include ecological solutions and not only grey infrastructure. An important alternative to the rainwater drainage system should include bioswales, rain gardens, and wetlands. When incorporated into the conventional grey system, these solutions can provide significant improvements. This is why the second workshop on Nature-Based Solutions was held, which was an important need for the city. Additionally, participatory process can facilitate sustainable development, promote public–private partnerships, and encourage local populations to engage with and care for the natural environment.

9.5.2 Challenge: Deforestation

The land use and land change map from 1985 to 2022 (Fig. 9.22) around the urban area reveals a significant transformation in the surrounding native vegetation. This is largely attributed to the suppression of native vegetation for the expansion of pastures and agriculture, particularly monoculture of soy and other temporary crops, but also by the urban growth. The urban expansion is evident in the northern and eastern regions of the original urban network and in some portions of the southern region. As the topography map demonstrated (Figs. 9.4 and 9.5), the city is constrained from expanding to the southern region due to the greater slope in this region. Conversely, in the northern region where the city has expanded, there is a substantial forest fragment.

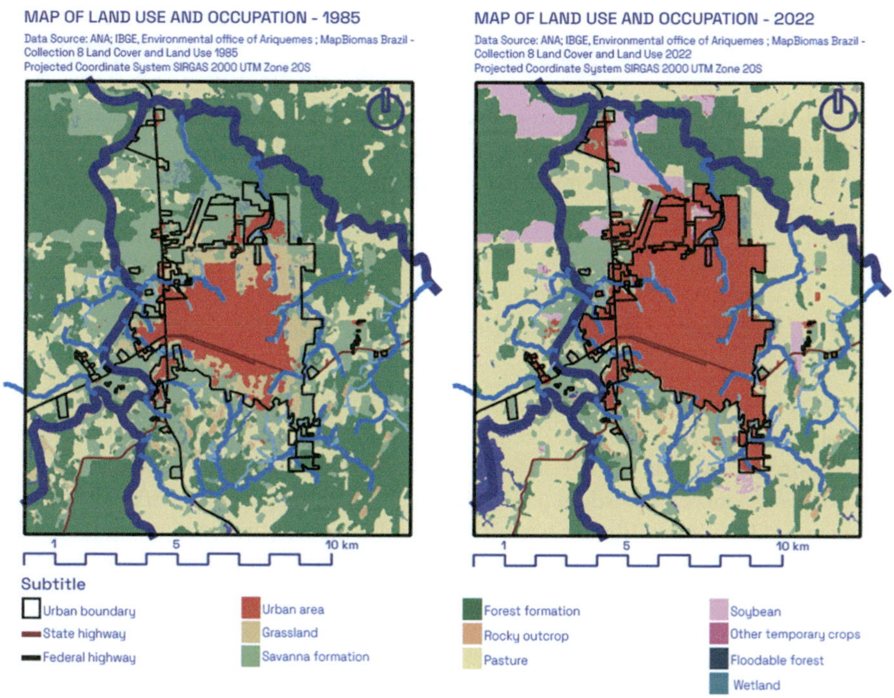

Fig. 9.22 Land use and land change analysis. (Source: Authors' own elaboration (Rocha))

Further insight was gained during the field visit regarding the challenge of forest maintenance. This was evidenced by the verification of urbanization pressure on the northern region of the city. Furthermore, the northern region has also experienced the greatest expansion of soybeans, which has occurred over savanna and forest areas.

Deforestation represents a significant challenge. Two main approaches to addressing this issue exist: the first is to restore forest areas, and the second is to avoid any further expansion of human activities in areas where native vegetation occurs naturally. This second strategy could be implemented in agribusiness and urban expansion. In the first case, the implementation of more advanced technology within the field of agribusiness could prove beneficial in reducing the reliance upon less sustainable practices. In the second instance, urban densification may offer a solution.

9.5.3 Challenge: Lack of Green Public Spaces (Squares and Parks)

The municipality of Ariquemes has established a system of green areas comprising squares, parks, gardens, and conservation units, as outlined in the Municipal Master Plan (Law 2341/2019). Three of the parks are designated as conservation areas. The Botanical Garden, which is situated within the urban fabric, encompasses 13.54 ha. The Ariquemes Municipal Park, located at the south end on the left side of BR-364, has a size of 151.07 ha. The Municipal Park of Canaã, situated at the south end on the right side of BR-364, is also a conservation area. The number of public squares identified was 14, while the number of parks was 5, including conservation units. This represents a relatively small number of publicly accessible green spaces, as evidenced by the Fig. 9.23.

9.5.4 Potential: The Institutional Sector and the Alleyways

The modernist planning of Ariquemes defined the central axis of the urban area as an Institutional sector, wherein open spaces play a significant role in terms of land use. Along this axis, there are collector roads that delineate the city's sectors. These sectors are organized according to a logic where services and commerce are concentrated around the institutional sector. The neighbourhoods are arranged in rectangular blocks that are linked in a regular manner, as illustrated in Fig. 9.7 on the Map of Avenues and Alamedas (De Paiva & Trevisan, 2023).

The institutional sector, in conjunction with the avenues and boulevards, has the potential to enhance the ecological quality of the area in a manner that is respectful of the natural Amazonian territory. In light of this perception, the potential of these environments is aligned with the implementation of practices that are in accordance

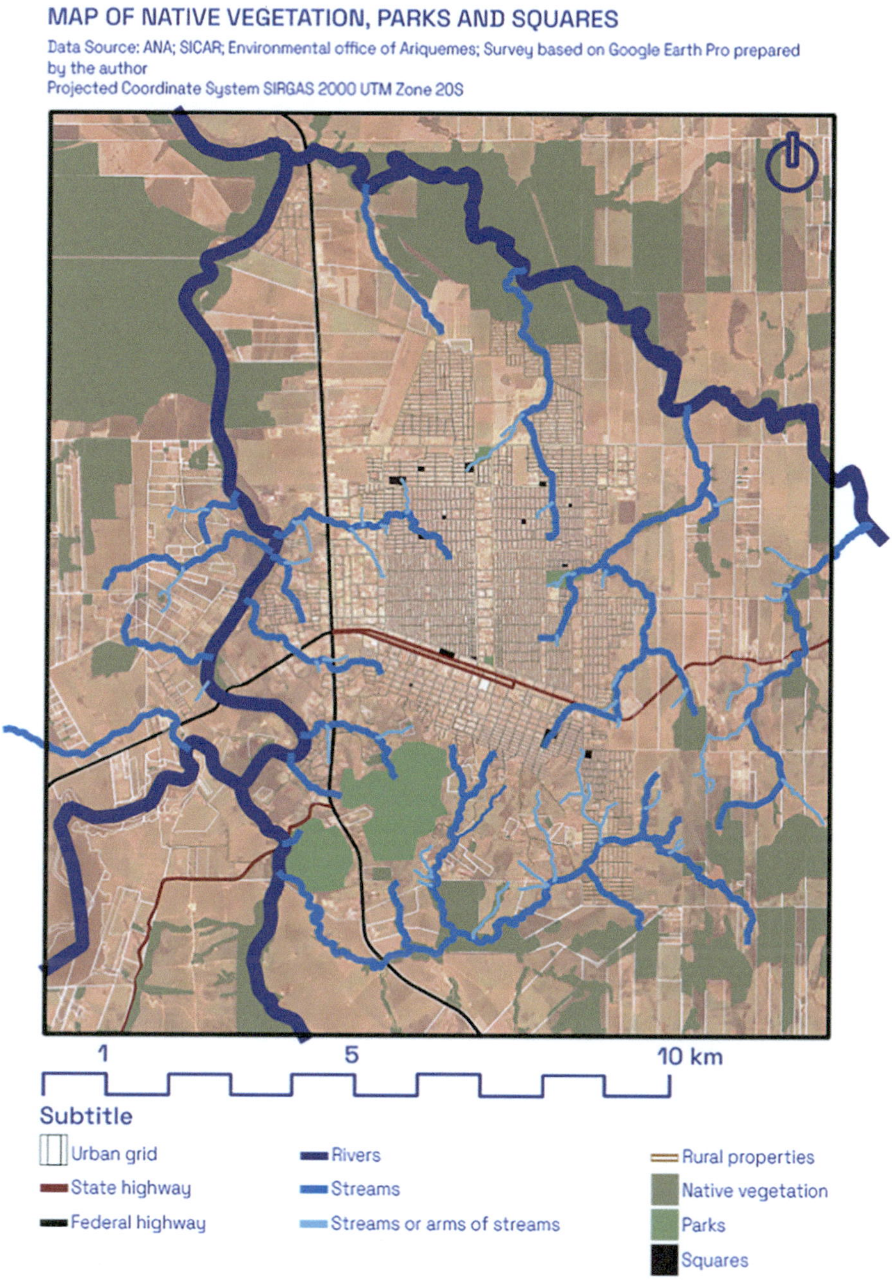

Fig. 9.23 Urban green areas system and native vegetation areas. (Source: Authors' own elaboration (Rocha))

with Nature-based Solutions and that seek to address issues that have arisen from the city's traditional and modern urban planning.

The alleyways, for example, which are 6 and 8 m wide, have the potential to facilitate connectivity between open spaces within the municipality. They can serve as green corridors, reintegrating the environment into the biome and providing greater well-being to the community. Furthermore, the area can also benefit from the increase in green areas.

9.5.5 *Challenge and Potential: The River System and the Vegetation Fragments*

The city of Ariquemes is situated within a complex river network, comprising the Rivers Canaã, Rio Branco, Rio Jamari, and 78 streams that originate or traverse the urban perimeter. This network encompasses a total length of 136 km (Kozerski, 2023). The river network serves as a primary system of natural drainage basins within Ariquemes (Ferreira, 2021).

In conjunction with the design of the river network, a system of green areas has been established by the Municipal Master Plan (Law 2341/2019, art. 188) as permanent preservation areas on the banks of watercourses. The minimum width of these areas is 100 m for the Jamari River, 50 m for the Canaã and Rio Branco Rivers, 30 m for the Igarapés, and 50 m for the springs (Kozerski, 2023). However, the streams that shape the urban fabric of Ariquemes have their banks inappropriately occupied by residential properties that discharge sewage and solid waste into the waterways. Such streams are frequently contaminated and degraded, lacking riparian forest coverage and exhibiting signs of siltation (Ferreira, 2021). As illustrated in the Figs. 9.24 and 9.25, a considerable number of residential properties are situated within areas identified as being at risk. Furthermore, it is evident that the municipal authorities have not upheld the requisite standards of protection in these designated zones.

In addition to the aforementioned areas, the municipality's neighbouring regions encompass a variety of forest fragments comprising native vegetation. Additionally, these fragments are present in non-buildable urban areas, including flooded regions, swamps, and riparian forests (Kozerski, 2023).

The city, as part of nature, should be planned with an eye on its ecosystemic composition (Meneguetti, 2007). However, urban expansion often fragments the landscape, annihilates biodiversity, and devastates habitats (Ndubisi, 2014).

Connecting open spaces, including soil and water not covered by buildings (Magnoli, 2006), helps keep urban centres in balance by protecting sensitive areas, providing habitats for wild animals, helping with urban drainage, and creating recreational environments. The number, quality, and arrangement of these spaces are key to the environmental functions of cities (Meneguetti, 2007). One way to connect these spaces is through nature-based solutions, like green corridors. These are strips of vegetation that connect green spaces in the urban core and add to the green spaces of a city (World Bank, 2021).

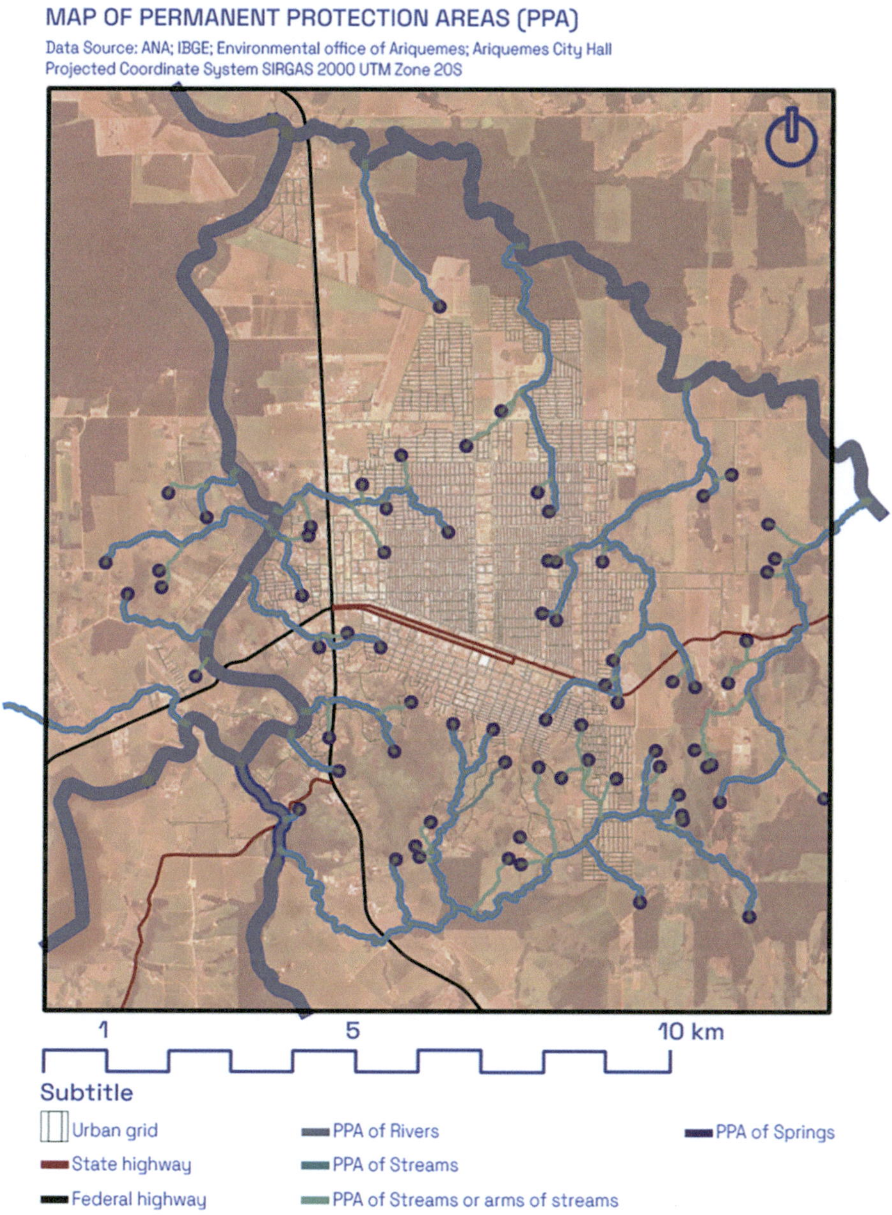

Fig. 9.24 Map of Permanent Protection Areas (PPA). (Source: Authors' own elaboration (Rocha))

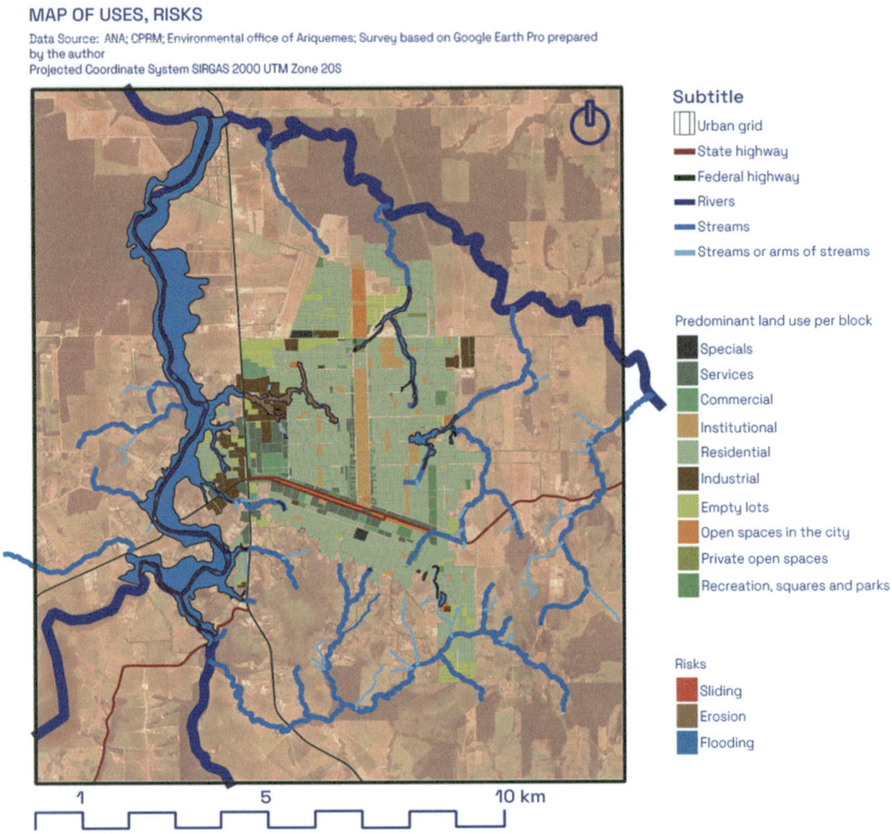

Fig. 9.25 Map of Land Use and risk areas. (Source: Authors' own elaboration (Rocha))

9.6 NbS Adaptation to the Amazon Reality: The SoN Concept

The concept of NbS was first mentioned by the World Bank in 2008. In 2015, it was officially introduced in the policy agenda of the European Commission, linked to the concept of renaturating cities. In other countries, alternative concepts were developed, such as the Sponge City in China. The concept emerged from working with nature, a concept that is not necessarily new, but together on the same "umbrella" of natural systems created an understandable view of the solutions and how to implement them at the urban scale, for example as a systemic approach for sustainable green drainage solutions.

Nature-Based Solutions are strategies that use natural processes to restore biodiversity, reduce the risk of catastrophes, promote water and food security, and improve human health. They can be done at three different scales: river basin, urban, and neighbourhood (World Bank, 2021). Whether any natural solution can be considered NbS is still not clear. The scientific community is constantly debating about NbS and which interventions could be classified as NbS (Sowińska-Świerkosz & García, 2022; Dumitru & Wendling, 2021).

The forest fragments, riparian forests, parks, squares, and other open spaces in the municipality of Ariquemes can be used to apply solutions at three different scales. This helps to connect urban green spaces with the forest fragments around them. Reconnections can also help with urban drainage, since they can increase water infiltration by implementing other solutions integrated into the corridors, such as bioswales, rain gardens, and retention areas.

It is evident that there is a need to enhance the relationship between urban areas and their natural or original surroundings. It is similarly evident that improvements to water drainage and treatment are required, as is an increase in green areas within urban areas. Nevertheless, it is pertinent to question whether the term "Nature-based Solutions" is an appropriate designation for these strategies in the context of the Amazon region. The rationale for this inquiry is rooted in the recognition that, in a region where numerous external (and frequently Eurocentric) perspectives have been imposed upon the cultural and original heritage, it is neither prudent nor appropriate to address the local issues with European solutions.

It is evident that NbS is a concept that originated in Europe but has since evolved to encompass a multitude of globally recognized solutions, including green roofs and wetlands. This shift has eased the burden for European researchers seeking to collaborate at the local level. Conversely, it would appear that the most reasonable course of action would be to give the concept another name that more accurately reflects the local culture and reality of different locations from the Global South.

The field trip and workshop were instrumental in facilitating our understanding that the concept of ecological thinking has been a fundamental aspect of Amazonian culture for centuries. However, the region's history and culture have been significantly impacted by various factors, including the destruction of the rainforest in Ariquemes and the extinction of the Arikêmes indigenous group. Moreover, it was largely erased and replaced by a culture of urban planning that is disconnected from the natural world and from the territory's history and natural characteristics.

As external researchers, introducing an external concept developed in Europe was a challenging and potentially dangerous situation. It was therefore necessary to adapt the NbS concept to the local reality. During the second workshop, an adaptation was made to the approach, but no new designation was created at that moment. The post-workshop reflections were instrumental in the development of this novel approach, designated as Solutions Originating from Nature (SoN).

The SoN conceptual approach for the Amazon region aims to enhance visibility of the historical and ongoing relationships between the original communities and the forest. These relationships have existed since time immemorial, yet have been largely erased in the occupation process or Rondônia by migrants.

The authors deemed it more respectful to devise a new designation for solutions that harness the potential of nature for climate mitigation and adaptation in the Amazon context than to merely adopt existing European terminology. Our future research in the region will continue in this vein.

9.7 Summary and Conclusions

The state of Rondônia had a history of forced solutions. First, the government encouraged migrants to settle in the forest region, despite the forest people already living there. These projects were the result of large-scale colonization in the 1970s. They occurred together with the creation of highways and roads that supported agribusiness expansion, causing deforestation.

This colonization process led to the construction of Ariquemes, designed by architect Antônio Carpinteiro. The city was planned according to modernist principles. These include using a regular urban grid and dividing land use into sectors. The city's name was inspired by the extinguish Arikêmes indigenous group, even though the history of the city does not honour the indigenous group.

The repeated impositions had a significant social and environmental impact on an urban area that has become detached from its natural environment and a society that has largely lost sight of its cultural roots. The researchers were eager to comprehend the demands of the social groups and to ascertain whether there was local interest and political support for the recovery of some lost boundaries. Fortunately, the researchers were able to identify not only interest but also numerous local initiatives developed by the local government and universities, especially related to recreating the environmental connection. This enabled the researchers to establish a partnership and implement two participatory processes, called co-planning workshops.

The transfer of academic or scientific knowledge to the local level could be optimized through the implementation of participatory processes. Moreover, the process benefited the researchers, as it enhanced the analytical complexity and facilitated the necessary adjustments to research outcomes. Furthermore, participatory processes enabled the integration of local knowledge into research. Two participatory workshops, held in Ariquemes, Rondônia, in 2023 and 2024, were presented here as illustrative cases of ecological co-planning workshops.

A participatory process is a complex method that requires careful consideration and planning to ensure its success. In a city in the Amazon region, which is characterized by a multifaceted blend of cultural, environmental, and societal elements, the actions of external actors at the local level can be particularly challenging. There is a potential for these actors to inadvertently impose ideas and solutions that may not align with the specific realities of the community. For this reason, it is essential to employ an attentive and respectful participatory method, which enables the construction of knowledge in collaboration with the participants, guiding them to reflect on their own reality.

The post-workshop activities served to consolidate some analysis and reflections about the actions and the local scale. Furthermore, these developments gave rise to novel conceptual frameworks for the region, reflecting the researchers' apprehensions regarding the tenuous boundary between guidance and intervention. This is the case of the concept "Solutions Originating from Nature (SoN), that adapt the NbS concept to this particular reality.

Conclusions of this experience can be summed-up in three important points: First, our findings indicate that it is imperative for researchers to collaborate with the local community to implement ecological co-planning processes for the landscape. This can facilitate a mutual transformation of values regarding urban and environmental design, thereby contributing to the construction and management of territories that prioritize the landscape, culture, heritage, and natural environment. This, in turn, reduces conflict between environmental preservation, urban development, and impositions from external points of view.

Secondly, SoN should be considered as an important tool for sustainability, starting from the local community and original knowledge to create the necessary solutions. This is crucial when considering the need to develop mitigation strategies in cities where an urban planning disconnected from nature has been developed, as in the case of Ariquemes.

Furthermore, our third conclusion is that there is still a lot of imposition in urban planning in the Global South. It is hoped that the external (European) solutions could be a good fit for the complex problems they face, and in this sense it's common to see the NbS concept exported to local realities, such as in Brazil, and not always adapted. SoN is now only a concept; it is a way to remind researchers and planners to respect the local cultural heritage. In other words, SoN offers a way for researchers to pay attention to the local scale and original knowledge when proposing sustainable solutions. The further development and implementation of SoN seems to be the best solution for Ariquemes and other Amazonian cities, and this idea will guide the authors in future research.

Acknowledgements The authors would like to express their gratitude to the Alexander von Humboldt Foundation (Georg Forster Experienced Researcher Grant to Verônica Garcia Donoso), UNIFAEMA University, and the Municipal Environmental Secretary of Ariquemes (SEMA) for their invaluable support during the workshops. The authors also gratefully acknowledge Dr. Mayumi Hirye, colleagues from the Chair and Institute of Urban Design (RWTH Aachen University), colleagues from QUAPÁ Brazilian Research Lab, professionals from Kanindé Ethno-environmental Defense Association, professionals from Porto Velho City Hall and Cacoal City Hall, teachers from the elementary schools of São Carlos do Jamari District. The authors would also like to extend their appreciation to all the participants of the workshops:

References

Ariquemes, P. M. (2019). *Lei Municipal n° 2.341 de 03 de julho de 2019: Plano Diretor Participativo*. Retrieved Jun 2024, from https://legislacao.camaradeariquemes.ro.gov.br/ver/FDDC381A/

Ariquemes, P. M. (2023). *Secretaria Municipal de Meio Ambiente - SEMA lança Projeto "RG dos Igarapés"*. Retrieved Mai 2024, from https://ariquemes.ro.gov.br/noticias/meio-ambiente/secretaria-municipal-de-meio-ambiente-sema-lanca-projeto-rg-dos-igarapes

Barifouse, L. (2013). O planejamento urbano entre ordem e o caos. *Revista De Direito Da Cidade., 5*(01), 146. https://doi.org/10.12957/rdc.2013.9727

Carcará, M. S. M., da Silva, E. A., & Moita, J. M. (2019). Saneamento básico como dignidade humana: entre o mínimo existencial e a reserva do possível. *Engenharia Sanitária e Ambiental, 24*, 493–500. https://doi.org/10.1590/S1413-41522019183905

Cavalcante, W. H. (2015). *Os Arikêmes e o SPI; o desafio da reelaboração cultural indígena sobre poder tutelar do estado Brasileiro*. Master's thesis of the academic master's program in history and cultural studies. Federal University of Rondônia, Porto Velho, RO, Brazil.

Cavalvante, W. H. (2016). A utilização da mídia escrita e da caravana ford como estratégias do governo de Rondônia para conclusão da BR-29. In O. Burgeile, J. L. P. Bueno, & J. M. Pacífico (Eds.), *Olhares da mídia na Amazônia: movimentos e manifestações*. EDUFRO.

Cordeiro, M. S. S. (2018). Pioneiros, fundadores e aventureiros—a ocupação de terras em Rondônia. *Revista de Antropologia, 61*(21), 125. https://doi.org/10.11606/2179-0892.ra.2018.145519

Costa, L. (1928). *O arranha-céu e o Rio de Janeiro. O Paiz*, 1 Jul. p. 4.

de Martins, J. S. (2019). A reinvenção da cidade na selva. *Tempo Social, 31*(1), 11–33. https://doi.org/10.11606/0103-2070.ts.2019.151225

De Paiva, L. N. Q., Trevisan, R. (2023). *Ariquemes (RO) e o arq. Antonio Carpintero: projeto e projetista de uma cidade nova em franjas de colonização*. in: XX ENANPUR, Belém. Retrieved April 2024, from https://anpur.org.br/anais-xxenanpur/

de Souza, J. A. O. (2020). Colonização na década de 1970, Rondônia e a BR-364. *Espaço em Revista, 22*(1), 82–100. https://doi.org/10.5216/er.v22i1.63286

Donoso, V. G. (2017). *Paisagem e cotidiano em habitação social nas regiões metropolitanas de São Paulo e Santiago do Chile*. PhD. Tesis, Faculty of Architecture and Urbanism, São Paulo University, São Paulo, Brazil. https://doi.org/10.11606/T.16.2018.tde-09062017-110211

Dumitru, A., & Wendling, L. (Eds.). (2021). Evaluating the impact of nature-based solutions: A handbook for practitioners. *European Commission EC*. https://doi.org/10.2777/244577

European Commission. (2015). *Towards an EU research and innovation policy agenda for nature-based solutions & renaturing cities*. European Commission.

Faria, R., Schvarsberg, B., & (Org.). (2011). *Políticas Urbanas e Regionais no Brasil*. University of Brasília, Faculty of Architecture and Urbanism.

Ferreira, C. C. F. (2021). *Caracterização ambiental dos igarapés da cidade de Ariquemes-Rondônia*. Course completion work in Environmental and Sanitary Engineering at FAEMA University. Ariquemes, 2021. Retrieved June 2024, from https://repositorio.unifaema.edu.br/bitstream/123456789/3028/1/C%C3%8DNTIA%20CARLA%20FERRAZO%20FERREIRA.pdf

Gehl, J. (2010). *Cities for people*. Island Press.

Gomes, E. T. (2019). *A colonização em Rondônia (1970–1980): estudo da atual configuração fundiária da área do PIC JI-Paraná. Master's thesis of the Postgraduate Program in History*. Federal University of Mato Grosso.

Gonçalves, M. F., Brandão, C. A., & Galvão, A. C. F. (2003). *Regiões e cidades, cidades nas regiões: o desafio urbano-regional*. Editora Unesp/Anpur.

Gonçalves, C. D. A. B., Pereira, M. N., & da Freitas, C. C. (2006). Análise do ambiente residencial urbano visando a inferência de população utilizando dados de sensoriamento remoto orbital de alta resolução. *Geografia, 31*(2), 371–402. Retrieved Mai 2024, from https://www.periodicos.rc.biblioteca.unesp.br/index.php/ageteo/article/view/1372/2063

IAS - Instituto de Água e Saneamento. (2020). *Ariquemes, RO*. Retrieved Mai 2024, from https://www.aguaesaneamento.org.br/municipios-e-saneamento/ro/ariquemes#:~:text=O%20QUE%20VOC%C3%8A%20PRECISA%20SABER%20SOBRE%20ARIQUEMES%20(RO)%3A&text=Possui%20fundo%20municipal%20de%20saneamento,n%C3%A3o%-20t%C3%AAm%20acesso%20%C3%A0%20%C3%A1gua

IBGE (n.d.) (no data) *Historical data. Dados históricos dos censos demográficos*. Retrieved Jun 2024, from https://memoria.ibge.gov.br/historia-do-ibge/historico-dos-censos/dados-historicos-dos-censos-demograficos.html

Jacobs, J. (2011). *Morte e vida das grandes cidades*. Martins Fontes.

Keppke, R. S. (2013). Repensando as anistias de uso e ocupação do solo em São Paulo. *Revista LABVERDE, 6*, 160–171. Retrieved Mai 2024, from https://www.revistas.usp.br/revistalabverde/article/view/61912/64743

Kozerski, G. R. (2023). *Espírito Santo, Alan Duarte do*. Urbanização e Meio Ambiente. Prefeitura Municipal de Ariquemes.

Macedo, S., Queiroga, E. F., Campos, A. C. A., Galender, F., & Custodio, V. (2018). *Os Sistemas de Espaços Livres e a Constituição da Esfera Pública Contemporânea no Brasil* (Vol. 1). Edusp.

Maciel, L. A. (1998). *A nação por um fio: caminhos, práticas e imagens da "Comissão Rondon"*. Edusc.

Magnoli, M. M. (2006). O parque no desenho urbano. *Paisagem Ambiente, 21*, 199–214. https://doi.org/10.11606/issn.2359-5361.v0i21p199-213

Meira, G. L. (2017). *O desconhecimento da história dos indígenas Arikêmes pela população do município de Ariquemes. Course completion in History*. Southern Frontier University.

Meneguetti, K. S. (2007). *De cidade-jardim a cidade sustentável: potencialidades para uma estrutura ecológica urbana em Maringá*. PhD. Tesis, Faculty of Architecture and Urbanism, São Paulo University, São Paulo, Brazil. https://doi.org/10.11606/T.16.2007.tde-19052010-105818

Morelato, A. A. H. N., Cardoso, A. C. D. (2020). *Madeira Mamoré Railway Company: O legado da companytown à cidade de Porto Velho, Rondônia*. XII SIIU, São Paulo e Lisboa. Retrieved Jun 2024, from https://revistes.upc.edu/index.php/SIIU/article/view/9852/1599

Ndubisi, F. O. (2014). *The ecological design and planning reader*. Island Press.

Pereira, E. M. (2017). Como Anda a participação? As condições Para a elaboração de planos diretores participativos | how is the participation going? The conditions for elaborating participative master plans. *Revista Brasileira De Estudos Urbanos E Regionais, 19*(2), 235. https://doi.org/10.22296/2317-1529.2017v19n2p235

Peres, R. B., & de Chiquito, E. A. (2012). Ordenamento territorial, meio ambiente e desenvolvimento regional: novas questões, possíveis articulações. *Revista Brasileira De Estudos Urbanos E Regionais, 14*(2), 71. https://doi.org/10.22296/2317-1529.2012v14n2p71

Rolnik, R. (2009). Democracia no fio da navalha: limites e possibilidades para a implementação de uma agenda de Reforma Urbana no Brasil. *Revista Brasileira De Estudos Urbanos E Regionais, 11*(2), 31. https://doi.org/10.22296/2317-1529.2009v11n2p31

Rossini, D., Forest, C.; Viadana, M. I. C. F. de. (2001). *Análise ambiental e do padrão espacial em áreas de expansão urbana de Limeira-SP*. Accessed Mai 2024, from http://www1.rc.unesp.br/igce/newpos/new_geo/downloads/2002/analise_ambiental.pdf

Secretaria de Estado do Desenvolvimento Ambiental. (2016). *Relatório final do PMSB de Ariquemes*. Mai 2016. Retrieved Jun 2024, from https://ariquemes.sedam.ro.gov.br/wp-content/uploads/2019/09/Plano-Municipal-de-Saneamento-Basico-Ariquemes.pdf

Souza, M. M. O. (2010). Entre a ferrovia do diabo e o trem fantasma: uma viagem pela história da estrada de ferro Madeira-Mamoré. Revista Campo-Território. *Revista da Geografia Agrária, 5*(9), 237–246. https://doi.org/10.14393/RCT5911934

Sowińska-Świerkosz, B., & García, J. (2022). What are nature-based solutions (NBS)? Setting core ideas for concept clarification. *Nature-Based Solutions, 2*, 1. https://doi.org/10.1016/j.nbsj.2022.100009

Tucci, C. E. (2003). Drenagem Urbana. *Ciência e cultura, 55*(4), 36–37. Retrieved April 2024, from http://cienciaecultura.bvs.br/scielo.php?script=sci_arttext&pid=S0009-67252003000400020

Work Bank. (2021). *A catalogue of nature-based solutions for urban resilience*. World Bank Group.

World Bank. (2008). *Climate change and adaptation: Nature-based from the World Bank portfolio*. Biodiversity.

Chapter 10
A Rethink of Urban Planning Towards an Ecological Planning Perspective

Verônica Garcia Donoso and Christa Reicher

Abstract This chapter offers concluding remarks on the subject of this book. It emphasizes the importance of integrating urban areas with natural environments as a basic tenet of urban design, noting that the degree of integration depends on various factors, including historical, political, environmental, and socioeconomic aspects. In addition, this chapter identifies shortcomings in some proposed designs and presents potential avenues for advancing ecological planning through participatory processes. The text then turns its attention to the necessity of increasing awareness of the reality of Amazonian cities in the state of Rondônia. This awareness should be employed to facilitate a rethinking of urban planning in the region.

Keywords Amazon rainforest · Rondônia · Ecological planning · Urban design

A fundamental principle of urban design is to understand and respect the capacity and performance of the natural environment, while integrating cities and nature. This principle is essential for building and handling environmental-friendly cities. Nevertheless, achieving a balance is variable and dependent on several factors, including historical, political, cultural, ecological, and socioeconomic aspects.

The Amazon region, a treasure for biodiversity and ecosystem services, has yet a big gap in this combination between its cities and nature. Many of the challenges have been explored within this book.

The difficulties faced in Rondônia have roots in history, as explored in Section 1 of this book. In addition to the effects of indigenous settlements dating back centuries, the drastic changes in its Land Use and Land Cover (LULC) from forests to

V. G. Donoso (✉)
Federal University of Santa Maria (UFSM), Cachoeira do Sul, Rio Grande do Sul, Brazil
e-mail: veronica.donoso@ufsm.br

C. Reicher
Institute for Urban Design and European Urbanism RWTH Aachen University, Aachen, Germany

V. G. Donoso, C. Reicher (eds.), *Rethinking Amazonian Cities in the State of Rondônia*, The Latin American Studies Book Series,
https://doi.org/10.1007/978-3-032-01817-5_10

cattle ranching and agriculture is linked with the arrival of settlers in the 1960s. The histories of the cities discussed in this book are also intertwined with this process. Government programmes to encourage people to move to the region, supported by the building of infrastructure (including roads, telecommunications, hydropower, and urban areas), have caused significant levels of deforestation. Once the roads had been constructed, logging, cattle farming, and agriculture became the most commonly associated activities with the loss of forested areas.

Furthermore, the rate of deforestation remains high, resulting in the degradation of habitats and loss of species. In addition, this process contributes to global climate change by releasing greenhouse gases (GHG) into the atmosphere.

The Amazon rainforest is a one-of-a-kind ecosystem on our planet. It is a valuable world heritage site due to its diverse plant and animal species and the environmental benefits it provides. Despite its significance, scientific understanding of the Amazon remains disproportionately low when compared to the vastness and importance of the area. Additionally, such scientific knowledge is yet to become necessary for informing decisions related to Land Use and development in the region, as well as supporting the preservation of its natural ecosystems' integrity.

Presenting evidence, examples and future methodologies for rethinking cities in the Amazon region is the main objective of this book. The authors of this book propose the use of nature within cities to address society's challenging issues of native flora and fauna depletion. Key solutions include green and blue urban areas, nature-based solutions (NbS) and other green infrastructure techniques that provide localized benefits to specific species or communities.

Some of the proposed solutions presented in part II of this book have limitations, as they were designed conceptually by authors without direct experience of the cities. Nevertheless, these studies aim to present a range of possible approaches to urban planning that are suited to the specificities of Amazonian cities. These can guide planners, landscape professionals, and local governments in comprehending the opportunities to restore cities and enhance urban development in a more combined and responsive manner.

Conversely, part III dedicated a chapter to examining the potential of actions at the local level. Chapter 9 presented challenges and possibilities in this context. The importance of participatory processes for integrating local knowledge and scientific research results was highlighted as was the importance of incorporating an ecological co-planning approach for cities in the Amazon region. The authors dedicated analysis and reflections to the results of their actions at the local level, developing a novel conceptual framework for the region, the "Solutions Originating from Nature" (SoN), which adapts the NbS concept to this reality, enhancing visibility to the original communities' knowledge.

The authors of this book have looked beyond old-style urban and landscape planning to explore the uniqueness of a diverse ecosystem. Their research shows how using sustainable techniques in our cities can enhance our health and protect biodiversity and ecosystem services.

Every chapter suggests that we raise greater awareness of how significant urban nature is to our residents. Given the crucial role played by the Amazon region in global climate, it is important to prioritize sustainable solutions and ecological interventions when planning Amazonian cities. These measures are essential and must be advocated and promoted to encourage a rethinking of urban planning in the region.

Index

V. G. Donoso, C. Reicher (eds.), *Rethinking Amazonian Cities in the State of Rondônia*, The Latin American Studies Book Series,
https://doi.org/10.1007/978-3-032-01817-5

MIX
Papier aus verantwortungsvollen Quellen
Paper from responsible sources
FSC® C105338

If you have any concerns about our products,
you can contact us on
ProductSafety@springernature.com

In case Publisher is established outside the EU,
the EU authorized representative is:
Springer Nature Customer Service Center GmbH
Europaplatz 3, 69115 Heidelberg, Germany

Printed by Libri Plureos GmbH
in Hamburg, Germany